MW00814536

Springer Series in
MATERIALS SCIENCE 86

Springer Series in
MATERIALS SCIENCE

Editors: R. Hull   R. M. Osgood, Jr.   J. Parisi   H. Warlimont

The Springer Series in Materials Science covers the complete spectrum of materials physics, including fundamental principles, physical properties, materials theory and design. Recognizing the increasing importance of materials science in future device technologies, the book titles in this series reflect the state-of-the-art in understanding and controlling the structure and properties of all important classes of materials.

Volumes 20–70 are listed at the end of the book.

S. Siebentritt   U. Rau   (Eds.)

Wide-Gap Chalcopyrites

With 122 Figures (1 color) and 19 Tables

 Springer

Dr. Susanne Siebentritt
Hahn-Meitner-Institut
Glienicker Str. 100, 14109 Berlin, Germany
Tel.: +49-30-8062 2442 fax: +49-30-8062 3199
E-mail: siebentritt@hmi.de
http://www.hmi.de/pubbin/vkart.pl?v=zky

Dr. Uwe Rau
Institute of Physical Electronics, University of Stuttgart
Pfaffenwaldring 47, 70569 Stuttgart, Germany
E-mail: uwe.rau@ipe.uni-stuttgart.de

Series Editors:

Professor Robert Hull
University of Virginia
Dept. of Materials Science and Engineering
Thornton Hall
Charlottesville, VA 22903-2442, USA

Professor R. M. Osgood, Jr.
Microelectronics Science Laboratory
Department of Electrical Engineering
Columbia University
Seeley W. Mudd Building
New York, NY 10027, USA

Professor Jürgen Parisi
Universität Oldenburg, Fachbereich Physik
Abt. Energie- und Halbleiterforschung
Carl-von-Ossietzky-Strasse 9–11
26129 Oldenburg, Germany

Professor Hans Warlimont
Institut für Festkörper-
und Werkstofforschung,
Helmholtzstrasse 20
01069 Dresden, Germany

ISSN   0933-033X

ISBN   3-540-24497-2 Springer Berlin Heidelberg New York

ISBN   978-3-540-24497-4 Springer Berlin Heidelberg New York

Library of Congress Control Number: 2005933718

This work is subject to copyright. All rights are reserved, whether the whole or part of the material is concerned, specifically the rights of translation, reprinting, reuse of illustrations, recitation, broadcasting, reproduction on microfilm or in any other way, and storage in data banks. Duplication of this publication or parts thereof is permitted only under the provisions of the German Copyright Law of September 9, 1965, in its current version, and permission for use must always be obtained from Springer. Violations are liable to prosecution under the German Copyright Law.

Springer is a part of Springer Science+Business Media.
springer.com

© Springer-Verlag Berlin Heidelberg 2006

The use of general descriptive names, registered names, trademarks, etc. in this publication does not imply, even in the absence of a specific statement, that such names are exempt from the relevant protective laws and regulations and therefore free for general use.

Typesetting by the Authors and SPI Publisher Services using a Springer TeX macro package
Cover concept: eStudio Calamar Steinen
Cover production: *design & production* GmbH, Heidelberg

Printed on acid-free paper   SPIN: 11382843   57/3141/SPI   5 4 3 2 1 0

Preface

Thin film solar modules are considered as the next generation of photovoltaics technology due to their higher cost reduction potential compared to conventional photovoltaic modules based on Si wafers. The cost advantages are due to lower material and energy consumption, lower semiconductor quality requirements, smaller dimensions of thin films and integrated module production, leading to reduced manpower needs. Currently, thin film technologies are boosted additionally by the shortage in the supply of silicon. Thin film modules based on Cu-chalcopyrite absorbers represent the most advanced thin film technology with high efficiency laboratory cells and mass production starting 2006. These high-efficiency cells and commercial modules are based on absorbers with a bandgap around 1.1 eV. In recent years, the interest in chalcopyrite absorbers with wider bandgaps has considerably increased due to the efforts to increase solar cell effiencies by using absorbers closer to the solar spectrum optimum and by constructing a thin film tandem cell. To date, solar cells based on wide gap chalcopyrites have failed to reach the excellent performance levels of their low gap "cousins." The authors of this book have set out to investigate the reasons behind the inferior behaviour of the widegap chalcopyrite solar cells and to suggest solutions. The chapters in this book address the various aspects of this question in analysing the properties of wide-gap and low-gap materials. Most of the results presented here were obtained within a network research project funded by the German Ministry of Research and Education (BMBF): "Hochspannungsnetz" (high voltage network), which was aimed at characterising the defect and interface behaviour, as well as grain boundary properties in wide-gap chalcopyrites. The results were presented in two workshops in the fall of 2002 and 2003 in the village of Triberg in Black Forest and in Castle Reichenow near Berlin, respectively. In addition to the contributions of the project partners, the present compilation also contains papers from "external experts" invited to these workshops. We are especially grateful to D. Cohen, W. Mnch, J. van Vechten and W. Walukiewicz for joining the workshops and for their contributions to this book, highlighting important new aspects and original work that will be helpful for our ongoing research efforts.

Berlin and Stuttgart,   Susanne Siebentritt
July 2005   Uwe Rau

Contents

List of Contributors

M. Albrecht
Institute of Microcharacterisation
Department of Materials Science
and Engineering
University of Erlangen-Nuremberg,
Cauerstr. 6
91058 Erlangen, Germany
albrecht@ww.uni-erlangen.de

G.H. Bauer
Faculty of Mathematics and Natural
Sciences
Carl von Ossietzky University
26111 Oldenburg, Germany
g.h.bauer@uni-oldenburg.de

J.D. Cohen
Department of Physics
University of Oregon
Eugene, OR 97403, USA
dcohen@OREGON.UOREGON.EDU

G. Hanna
Institute of Physical Electronics
(IPE)
University Stuttgart
Pfaffenwaldring 47
70569 Stuttgart, Germany
george.hanna@zsw-bw.de

J.T. Heath
Department of Physics
Linfield College
McMinnville
OR 97128, USA
jheath@linfield.edu

A. Klein
Surface Science Division
Institute of Materials Science
Darmstadt University of Technology
Petersenstrasse 23
D-64287 Darmstadt, Germany
aklein@surface.tu-darmstadt.de

R. Kniese
Zentrum für Sonnenenergie- und
Wasserstoff-Forschung
Baden-Württemberg (ZSW)
Industriestr. 6
70565 Stuttgart, Germany
robert.kniese@zsw-bw.de

R. Menner
Zentrum für Sonnenenergie- und
Wasserstoff-Forschung
Baden-Württemberg (ZSW)
Industriestr. 6
70565 Stuttgart, Germany
richard.menner@zsw-bw.de

W. Mönch
Department of Physics
Universität Duisburg-Essen
47048 Duisburg, Germany
w.moench@uni-duisburg.de

N. Ott
Institute of Microcharacterisation
Department of Materials Science
and Engineering
University of Erlangen-Nurember,
Cauerstr. 6,
91058 Erlangen, Germany
Niels.Ott@ww.uni-erlangen.de

M. Powalla
Zentrum für Sonnenenergie- und
Wasserstoff-Forschung
Baden-Württemberg (ZSW),
Industriestr. 6 70565 Stuttgart,
Germany
michael.powalla@zsw-bw.de

U. Rau
Institut für Physikalische Elektronik
(IPE)
University of Stuttgart,
Pfaffenwaldring 47
70569 Stuttgart, Germany
uwe.rau@ipe.uni-stuttgart.de

S. Sadewasser
Hahn-Meitner Institut
Glienicker Str. 100
14109 Berlin, Germany
sadewasser@hmi.de

T. Schulmeyer
Surface Science Division
Institute of Materials Science
Darmstadt University of Technology,
Petersenstrasse 23
D-64287 Darmstadt, Germany
tschulmeyer@
surface.tu-darmstadt.de

W.N. Shafarman
Institute of Energy Conversion
University of Delaware
Newark, DE 19716, USA
wns@udel.edu

S. Siebentritt
Hahn-Meitner-Institut
Glienicker Str. 100
14109 Berlin, Germany
siebentritt@hmi.de

U. Stein
Zentrum für Sonnenenergie- und
Wasserstoff-Forschung
Baden-Württemberg (ZSW),
Industriestr. 6 70565 Stuttgart,
Germany
ulrike.stein@zsw-bw.de

H.P. Strunk
Institute of Microcharacterisation
Department of Materials Science
and Engineering
University of Erlangen-Nuremberg,
Cauerstr. 6
91058 Erlangen, Germany
strunk@ww.uni-erlangen.de

M. Turcu
Institute of Physical Electronics
University of Stuttgart
Pfaffenwaldring 47
70569 Stuttgart, Germany
mircea.turcu@ipe.uni-stuttgart.de

J.A. Van Vechten
School of Electrical Engineering and
Computer Science
Oregon State University
Corvallis, OR 97331-3211, USA
JAvanvec@msn.com

G. Voorwinden
Zentrum für Sonnenenergie- und
Wasserstoff-Forschung
Baden-Württemberg (ZSW),
Industriestr. 6 70565 Stuttgart,
Germany
georg.voorwinden@zsw-bw.de

W. Walukiewicz
Materials Sciences Division
Lawrence Berkeley National
Laboratory
MS 2R0200, 1 Cyclotron Rd.
Berkeley, CA 94720-8197, USA
W_Walukiewicz@lbl.gov

1

Cu-Chalcopyrites – Unique Materials for Thin-Film Solar Cells

S. Siebentritt and U. Rau

Thin-film solar modules are considered as the next generation of photovoltaics technology due to their higher cost-reduction potential compared to conventional photovoltaic modules based on Si wafers. The cost advantages are due to lower material and energy consumption, lower semiconductor quality requirements, short distances in thin films, and integrated module production leading to reduced manpower needs. Currently, thin-film technologies are boosted additionally by shortage in the supply of Si.

Thin-film modules based on Cu-chalcopyrite absorbers [1] represent the most advanced thin-film technology with laboratory cells reaching efficiencies above 19% [2]. Modules of $Cu(In,Ga)(S,Se)_2$ solar cells are in the pilot production stage at several places worldwide and large modules have reached efficiencies above 13% [3] and output power of 80 W [4]. Mass production in Europe will start in 2006 [5].

The basic structure of these solar cells and the schematics of their band structure are shown in Fig. 1.1. The p/n junction is formed between the p-type chalcopyrite absorber and the window layer, usually a double layer of undoped-ZnO and Al-doped or Ga-doped ZnO. The quality of the heterojunction is greatly improved by the introduction of a CdS buffer layer; alternative Cd-free materials are under investigation [6].

The following list gives a brief account of potentially critical points of the $Cu(In,Ga)(S,Se)_2$ thin-film solar cell technology and the research that is aimed to solve these issues:

1. The photovoltaic junction is made by a heterocontact between two non-lattice matched materials, a situation that is potentially hazardous because of interface recombination via a high density of interface states. It is therefore desirable that the Fermi level at the absorber/buffer interface is above midgap, i.e., the interface should be type-inverted with respect to the absorber [7]. However, despite these type-inversion efficiencies, close to 20% would remain out of reach with this type of heterojunction, unless the special feature of a Cu-poor layer that forms spontaneously on the

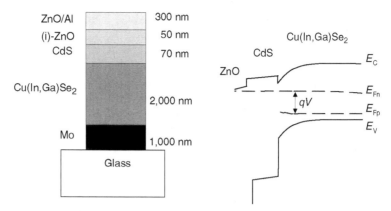

Fig. 1.1. Basic layer structure and energy-band diagram of a ZnO/CdS/Cu(In,Ga)Se₂ heterojunction solar cell under a bias with voltage V

surface of the absorber material would suppress interface recombination further as discussed in Chap. 6.

2. The Mo-coated glass serves as the substrate for the growth of the absorber material and, in addition, as the ohmic back contact for the completed solar cell. The glass substrate, usually ordinary soda-lime glass, is not a natural choice as part of a highly sophisticated electronic device, especially because of impurities, like alkali atoms contained in this material. Fortunately, introduction of Na from the soda-lime glass substrate contributes positively to the photovoltaic quality of the absorber material. Although the precise origin of this beneficial effect is not yet entirely understood, the presence of Na during absorber growth is mandatory for high-efficiency devices (for a detailed discussion see [8]). The suitability of Mo as the back contact of the device is also somewhat stunning as Mo on p-type CuInSe₂ is known to form a Schottky contact with a barrier height of 0.8 eV [9]. Hence, one would not guess that Mo does perform especially well as an ohmic back contact for a CuInSe₂ solar cell. Fortunately, it turns out that a MoSe₂ film of thickness of few ten nm forms on top of the Mo layer during absorber growth, thus enabling excellent ohmic properties of the back contact [10].

3. The Cu(In,Ga)(S,Se)₂ absorber is a polycrystalline material with a grain size between few hundred nm and few μm. Thus, the grain size barely matches the film thickness and is more than four orders of magnitude smaller than the size of the grains in polycrystalline silicon that delivers about the same solar cell efficiencies as the small-grained Cu(In,Ga)(S,Se)₂. Therefore, the electronic activity of grain boundaries and other extended defects must be extraordinarily low. Due to their benign manner, these defects seemed unimportant and only recently has attention been focused

on them. Possibly, the peculiar phase and defect behavior of the Cu-chalcopyrites material also influences the optoelectronic properties of grain boundaries as it does for the surfaces. The nature of grain boundaries is dealt with in Chaps. 9 and 10 .

4. The $ZnO/CdS/Cu(In,Ga)(S,Se)_2/Mo$ heterojunction solar cell is a very complicated system, with at least 11 chemical elements known to contribute actively to the electronic quality of the device. Because of the large variety of chemical reactions among these elements, chemical stability, especially at material interfaces, becomes an extremely critical question (for a review of the chemical stability, see [11]). Up to this point, we have dealt with the complex properties of the external and internal *interfaces* of the layer system shown in Fig. 1.1. Concentrating on the *bulk* properties of Cu-chalcopyrite absorbers, we have to consider the defect physics of a ternary compound.

5. A ternary compound (like $CuInSe_2$) already possesses 12 possible *intrinsic defects.* Therefore, even without considering the effects of In/Ga or S/Se alloying, we see that we are already dealing with a very complex defect chemistry. In addition, most of these intrinsic defects have very low defect formation energies, below $1\,eV$ [12], which is very small compared to any other compound or elemental semiconductor. Moreover, the formation energy of *defect complexes* of the structure $(2\,V_{Cu}-In_{Cu^{2+}})$ becomes negative in Cu-poor materials. The fact that this defect complex is electrically inactive is one of the reasons that make $CuInSe_2$ a photovoltaic-attractive material. The occurrence of several Cu-poor phases, e.g., $CuIn_3Se_5$ and $CuIn_5Se_8$, is explained by the regular ordering of these complexes. The electronically beneficial nature of these phases in solar cells results from the fact that they have a band-gap energy larger than the stoichiometric chalcopyrite. The formation of a surface layer with reduced Cu concentration at the surface of Cu-poor films (see point 1 and [13, 14]) plays a critical role in the interface formation as is discussed in Chaps. 6 and 11.

6. The ease with which defects are formed in chalcopyrite materials also leads to an unusually large existence region of the chalcopyrite phase, which extends far into the Cu-poor region [15, 16]. On the other hand, Cu excess is not incorporated into the chalcopyrite; it forms an additional Cu–Se phase at the surface.

7. The low defect formation energies are also the basis for the unusual stabilization of the polar surfaces. In other compound semiconductors like ZnSe and GaAs, the non-polar surfaces are the stable ones, while in chalcopyrites it is the (112) surfaces that are stable [17] – these correspond to the (111) surfaces in the cubic lattice. This can be explained by the removal of the surface dipole by defect formation [18, 19], leading to Cu-poor surfaces.

8. Maybe the major effect of the low defect formation energies is the fact that the Cu-chalcopyrites are doped by their native defects. No impurity doping is used to obtain the p-type nature of the solar cell absorbers.

Up to date, there exists no reliable information on the chemical nature
of the defects. Nevertheless, some trends are observed: CuInSe$_2$ tends to
be p-type under Se-excess or Cu-excess conditions and n-type under Se-
deficient conditions [20,21]. On the other hand, CuGaSe$_2$, which shows the
same shallow defects as CuInSe$_2$ (discussed in Chap. 7), is always p-type
under any stoichiometric deviation. The difficulty in obtaining chemical
information on the involved defects is directly based on the ternary nature
of these compounds. In binary materials, defect-chemical information can
be obtained from annealing experiments controlling the concentration of
one of the compounds. For ternary compounds this would only be possible by controlling the concentration of two of the three compounds [22],
which is experimentally extremely difficult. Therefore, most annealing experiments are done controlling only one component, rendering their results
very unreliable.

At this point, the reader is reminded that all beneficial circumstances discussed earlier are strictly valid only for a narrow composition range of the
Cu(In,Ga)(S,Se)$_2$ system, namely those alloys with low Ga and S contents.
Consequently, record-efficiency cells have a band gap energy, E_g of about 1.1–
1.15 eV. However, the entire alloy system allows the control of the band gap
between 1.05 eV(CuInSe$_2$) and 2.5 eV (CuGaS$_2$). Therefore, a considerable
amount of research and technological development directs towards those compositions that have a wider band gap than the up-to-now optimum material.
This is motivated by the following reasons [23, 24]:

1. The band gap energy $E_g \approx 1.1$ eV is below the optimum match to the
   solar spectrum. Therefore, higher efficiencies are expected from wider gap
   alloys as long as the recombination and transport properties of the wide-
   gap devices correspond to those of the best low-gap devices.
2. The lower current densities in wide-gap devices lead to lower resistive
   losses, thus allowing for wider cells within a solar module and thus reducing the number of necessary scribes for monolithic integration of the
   cells into a module. Also, the thickness of front and back electrodes can
   be reduced.
3. Wide-gap solar cells are expected to have a better temperature coefficient
   and, therefore, to perform better under real-world operating conditions
   than low-gap cells.
4. Wide-gap absorbers are better suited for space applications since the
   degradation of the open circuit voltage (V_{OC}) due to radiation is less
   critical in devices with a high V_{OC} than in low-gap devices with a corresponding low V_{OC} [25].
5. The free electron absorption in highly conducting ZnO window materials
   is not as critical for wide-gap materials as for low-gap materials, where
   the absorption of infrared light in ZnO overlaps with the absorption of
   the absorber material.

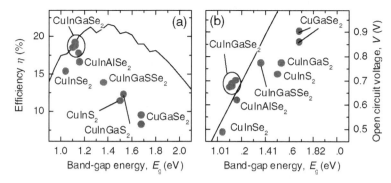

Fig. 1.2. (a) Highest published conversion efficiencies of solar cells from CuInSe$_2$ [4], Cu(In,Ga)Se$_2$ [2, 29, 30], Cu(In,Al)Se$_2$ [31], Cu(In,Ga)(S,Se)$_2$ [32], CuInS$_2$ [33], Cu(In,Ga)S$_2$ [34], and CuGaSe$_2$ [27,35] as a function of the band-gap energies E_g of the absorber materials. (b) Open circuit voltages of the cells shown in (a). The *solid lines* in (a) and (b) stem from an extrapolation of the recombination properties of the best Cu(In,Ga)Se$_2$ solar cells toward higher and lower E_gs

6. The wide range of band gap energies (E_g) of the Cu(In,Ga)(S,Se)$_2$ alloy system embraces combinations of E_g that in principle allow us to build Cu-chalcopyrite based tandem solar cells [26].

Thus, research and development on wide-gap chalcopyrite is an attractive issue because of the technological flexibility that is provided by mastering an alloy system with band gap energies matching the entire solar spectrum.

Unfortunately, all attempts to achieve efficiencies in the range of 17–20% by using wide-gap chalcopyrites have failed so far. For example, CuGaSe$_2$ solar cells have not reached the 10% efficiency level yet [27]; only by the addition of a small amount of In an efficiency of 10.5% was reached [28]. Figure 1.2a gives an overview of the power conversion efficiencies that have been obtained with different Cu-chalcopyrite alloys featuring a sharp drop when using band gap energies in excess of 1.2 eV. The reason for the low efficiencies lies in the fact that the open circuit voltages of these devices do not correspond to those values that are expected from their larger band-gap energies (Fig. 1.2b).

Up to now, research and development have not identified the physical origin of the relatively poor photovoltaic performance of the wide-gap chalcopyrites. Obviously, the unique solutions and features found while developing the low-gap absorbers are not directly suitable for their wide-gap counterparts. In addition, critical issues (see the eight-point list discussed earlier) have been solved and understood for the low-gap alloys after a long research effort. For their wide-gap counterparts, we should be aware that we may not necessarily benefit again from the goodwill of nature as much as in the low-gap alloys.

Only slightly deteriorated interface properties at the frontelectrode and back electrode, only slightly altered grain boundary properties and only comparatively small changes in the defect chemistry, may sum up to the considerably degraded performance. Thus, all ingredients that are obviously necessary to achieve the excellent photovoltaic performance of the low-gap chalcopyrites must be critically checked for their applicability to wide-gap alloys.

The following chapters address these different aspects in analyzing the properties of wide-gap and low-gap materials. Most of the results presented here were obtained in a network research project funded by the German Ministry of Research and Education (BMBF) – "Hochspannungsnetz" (high-voltage network), which was aimed at characterizing the defect and interface behavior as well as grain boundary properties in wide-gap chalcopyrites. The results were presented in two workshops in the fall of 2002 and 2003 in the village of Triberg in Black Forest and in castle Reichenow near Berlin, respectively. In addition to the contributions of the project partners, the present compilation also contains papers from "external experts" invited to these workshops. We are especially grateful to D. Cohen, W. Mönch, J. van Vechten, and W. Walukiewicz for joining the workshops and for their contributions to this book, highlighting important new aspects and original work that will be helpful for our ongoing research efforts.

References

1. Shay, J.L., Wernick, J.H.: Ternary Chalcopyrite Semiconductors: Growth, Electronic Properties and Application. Pergamon Press, Oxford (1975)
2. Ramanathan, K., Contreras, M.A., Perkins, C.L., Asher, S., Hasoon, F.S., Keane, J., Young, D., Romero, M., Metzger, W., Noufi, R., Ward, J., Duda, A.: Properties of 19.2% efficiency ZnO/CdS/CuInGaSe$_2$ thin-film solar cells. Prog. Photovolt. Res. Appl. 11, 225–230 (2003)
3. Probst, V., Stetter, W., Palm, J., Toelle, R., Visbeck, S., Calwer, H., Niesen, T., Vogt, H., Hernández, O., Wendl, M., Karg, F.H.: CIGSSE module pilot processing: from fundamental investigations to advanced performance. In: Kurokawa, K., Kazmerski, L.L., McNelis, B., Yamaguchi, M., Wronski, C., Sinke, W.C. (eds.) Proceedings of the 3rd World Conference on Photovoltaic Solar Energy Conversion, pp. 329–334. Arisumi Printing Inc., Osaka, Japan (2003)
4. Powalla, M., Dimmler, B.: New developments in CIGS thin-film solar cell technology. In: Kurokawa, K., Kazmerski, L.L., McNelis, B., Yamaguchi, M., Wronski, C., Sinke, W.C. (eds.) Proceedings of the 3rd World Conference on Photovoltaic Solar Energy Conversion, pp. 313–318. Arisumi Printing Inc., Osaka, Japan (2003)
5. Powalla, M., Dimmler, B., Schäffler, R., Voorwinden, G., Stein, U., Mohring, H.-D., Kessler, F., Hariskos, D.: CIGS solar modules – progress in pilot production, new developments and applications. In: Ossenbrinck, H.A. (ed.) Proceedings of the 19th European Photovoltaic Solar Energy Conference. pp. 1663–1668. JRC, Ispra, Italy (2004)
6. Siebentritt, S.: Alternative buffers for chalcopyrite solar cells. Solar Energy 77, 767–775 (2004)
7. Klenk, R.: Characterization and modelling of chalcopyrite solar cells. Thin Solid Films 387, 135–140 (2001)

8. Rau, U., Schock, H.W.: Electronic properties of Cu(In,Ga)Se$_2$ heterojunctio solar cells — recent achievements, current understanding, and future challenges. Appl. Phys. A 69, 131–147 (1999)
9. Jaegermann, W., Löher, T., Pettenkofer, C.: Surface properties of chalcopyrite semiconductors. Cryst. Res. Technol. 31, 273–280 (1996)
10. Nishiwaki, S., Kohara, N., Negami, T., Wada, T.: MoSe$_2$ layer formation at Cu(In,Ga)Se$_2$/Mo interfaces in high efficiency Cu(In$_{1-x}$Ga$_x$)Se$_2$ solar cells. Jpn. J. Appl. Phys. (Part 2) 37, 71–73 (1998)
11. Guillemoles, J.F., Kronik, L., Cahen, D., Rau, U., Jasenek, A., Schock, H.W.: Stability issues of Cu(In,Ga)Se$_2$-based solar cells. J. Phys. Chem. B 104, 4849–4862 (2000)
12. Zhang, S.B., Wei, S.-H., Zunger, A., Katayama-Yoshida, H.: Defect physics of the CuInSe$_2$ chalcopyrite semiconductor. Phys. Rev. (B) 57, 9642–9656 (1998)
13. Schmid, D., Ruckh, M., Grunwald, F., Schock, H.W.: Chalcopyrite/defect chalcopyrite heterojunctions on the basis of CuInSe$_2$. J. Appl. Phys. 73, 2902–2909 (1993)
14. Schmid, D., Ruckh, M., Schock, H.W.: Photoemission studies on Cu(In,Ga)Se$_2$ thin films and related binary selenides. Appl. Surf. Sci. 103, 409–429 (1996)
15. Fearheiley, M.: The phase relations in the Cu, In, Se system and the growth of CuInSe$_2$ single crystals. Solar Cells 16, 91–100 (1986)
16. Mikkelsen, J.C. Jr.: Ternary phase relations of the chalcopyrite compound CuGaSe$_2$. J. Electron. Mater. 10, 541–558 (1981)
17. Liao, D., Rockett, A.: Epitaxial growth of Cu(In,Ga)Se$_2$ on GaAs(110). J. Appl. Phys. 91, 1978–1983 (2002)
18. Jaffe, J.E., Zunger, A.: Defect-induced nonpolar-to-polar transition at the surface of chalcopyrite semiconductors. Phys. Rev. (B) 64, 241301–241304 (2001)
19. Zhang, S.B., Wei, S.-H.: Reconstruction and energetics of the polar (112) and (−1−1−2) versus the nonpolar (220) surfaces of CuInSe$_2$. Phys. Rev. (B) 65, 081402-1–081401-4 (2002)
20. Tell, B., Shay, J.L., Kasper, H.M.: Room-temperature electrical properties of ten I–III–VI2 semiconductors. J. Appl. Phys. 43, 2469 (1972)
21. Noufi, R., Axton, R., Herrington, C., Deb, S.K.: Electronic properties vs composition of thin films of CuInSe$_2$. Appl. Phys. Lett. 45, 668–670 (1984)
22. Bardeleben, H.J.V.: The chemistry of structural defects in CuInSe$_2$. Solar Cells 16, 381–390 (1986)
23. Siebentritt, S.: Wide gap chalcopyrites: material properties and solar cells. Thin Solid Films 403–404, 1–8 (2002)
24. Rau, U.: Electronic properties of wide-gap Cu(In,Ga)(S,Se)$_2$ solar cells. In: Kurokawa, K., Kazmerski, L.L., McNelis, B., Yamaguchi, M., Wronski, C., Sinke, W.C. (eds.) Proceedings of the 3rd World Conference on Photovoltaic Solar Energy Conversion, pp. 2847–2852. Arisumi Printing Inc., Osaka, Japan (2003)
25. Jasenek, A., Hahn, T., Schmidt, M., Weinert, K., Wimbor, M., Hanna, G., Orgassa, K., Hartmann, M., Schock, H.W., Rau, U., Werner, J.H., Schattat, B., Kraft, S., Schmid, K.H., Bolse, W., Roche, G.L., Robben, A., Bogus, K.: Stability of Cu(In,Ga)Se$_2$ thin film solar cells under 1 MeV electron radiation. In: Proceedings of the 16th European Photovoltaic Solar Energy Conference and Exhibition, pp. 982–985, 2000

26. Nishiwaki, S., Siebentritt, S., Walk, P., Lux-Steiner, M.C.: A stacked chalcopyrite thin-film tandem solar cell with 1.2 V open-circuit voltage. Prog. Photovolt. Res. Appl. 11, 243–248 (2003)

27. Young, D.L., Keane, J., Duda, A., AbuShama, J.A.M., Perkins, C.L., Romero, M., Noufi, R.: Improved performance in $ZnO/CdS/CuGaSe_2$ thin-film solar cells. Prog. Photovolt. Res. Appl. 11, 535–541 (2003)

28. Symco-Davies, M., Noufi, R., Coutts, T.J.: Progress in high performance PV-polycrystalline thin film tandem cells. In: Hoffmann, W., Bal, J.-L., Ossenbrinck, H.A. (eds.) Proceedings of the 19th European Photovoltaic Solar Energy Conference. pp. 1651–1656. JRC, Ispra, Italy (2004)

29. Orgassa, K., Nguyen, Q., Kötschau, I.M., Rau, U., Schock, H.W., Werner, J.H.: Optimized reflection of CdS/ZnO window layers in $Cu(In,Ga)Se_2$ thin-film solar cells. In: McNelis, B., Palz, W., Ossenbrinck, H.A., Helm, P. (eds.) Proceedings of the 17th European Photovoltaic Solar Energy Conference, WIP-Munich Germany, pp. 1039–1042, 2002

30. Negami, T., Hashimoto, Y., Nishiwaki, S.: $Cu(In,Ga)Se_2$ thin-film solar cells with an efficiency of 18%. Sol. Energy Mater. Sol. Cells 67, 331–335 (2001)

31. Marsillac, S., Paulson, P.D., Haimbodi, M.W., Birkmire, R.W., Shafarman, W.N.: High-efficiency solar cells based on $Cu(InAl)Se_2$ thin films. Appl. Phys. Lett. 81, 1350–1352 (2002)

32. Friedlmeier, T.M., Schock, H.W.: Improved voltages and efficiencies in $Cu(In,Ga)(S,Se)_2$ solar cells. In: Schmid, J., Ossenbrink, H.A., Helm, P., Ehmann, H., Dunlop, E.D. (eds.) Proceedings of the 2nd World Conference on Photovoltaic Solar Energy Conversion. European Commission, Luxembourg, pp. 1117–1120 (1998)

33. Siemer, K., Klaer, J., Luck, I., Bruns, J., Klenk, R., Braunig, D.: Efficient $CuInS_2$ solar cells from a rapid thermal process (RTP). Sol. Energy Mater. Sol. Cells 67, 159–166 (2001)

34. Kaigawa, R., Neisser, A., Klenk, R., Lux-Steiner, M.Ch.: Improved performance of thin film solar cells based on $Cu(In,Ga)S_2$. Thin Solid Films 415, 266–271 (2002)

35. Nadenau, V., Hariskos, D., Schock, H.W.: $CuGaSe_2$ thin film solar cells with improved performance. In: Ossenbrink, H.A., Helm, P., Ehmann, H. (eds.) Proceedings of the 14th European Photovoltaic Solar Energy Conference, Stephens, Bedford, pp. 1250–1253, 1997

2

Band-Structure Lineup at I–III–VI$_2$ Schottky Contacts and Heterostructures

W. Mönch

2.1 Introduction

As with all other semiconductor devices, the band-structure lineup in chalcopyrite solar cells also determines their electronic properties. For improvements of the fabrication processes and the design of new device concepts, it is desirable to have some insight into the physical mechanisms that determine the barrier heights and the band-edge offsets of the I–III–VI$_2$ Schottky contacts and heterostructures, respectively. As this chapter will demonstrate, the I–III–VI$_2$ chalcopyrites behave quite the same as all other semiconductors, in that their Schottky barrier heights and heterostructure-band offsets are also explained by the continuum of interface-induced gap states(IFIGS).

The rectifying properties of metal–semiconductor contacts were discovered by Braun [1]; and Schottky [2] explained them by a depletion layer on their semiconductor side. Schottky's explanation shifted the focus to the physical mechanisms, which determine the barrier heights of metal–semiconductor contacts, i.e., the energy position of the Fermi level, relative to the band-edge of the majority charge carriers at the interface. The early Schottky–Mott rule [3, 4] proposed the n-type (p-type) barrier height $\Phi_{\mathrm{Bn,p}}$ of a metal–semiconductor contact to equal the difference between the work function Φ_{m} of the metal and the electron affinity (ionization energy) of the semiconductor in contact. However, the slope parameter $S_{\Phi} = -\mathrm{d}\Phi_{\mathrm{Bp}}/\mathrm{d}\Phi_{\mathrm{m}}$ of metal–selenium rectifiers turned out to be much smaller than unity, the value predicted by the Schottky–Mott rule. Schottky [4] consequently concluded the failure of this simple rule in 1940. But most surprisingly, some groups still believe it to be valid for ideal Schottky contacts.

Bardeen [5] proposed electronic interface states to exist in the semiconductor band-gap at Schottky contacts. The charge absorbed in these interface states and the depletion layer then compensates the charge on the metal side of Schottky contacts and, as a consequence, the slope parameter S_{Φ} will become smaller than unity, the value predicted by the Schottky–Mott rule. Considering the quantum-mechanical tunnel effect at metal–semiconductor interfaces,

Heine [6] noted that for energies in the semiconductor band-gap the volume states of the metal have tails in the semiconductor. Tejedor and Flores [7] applied this idea to semiconductor heterostructures, where for energies in the band-edge discontinuities the volume states of one semiconductor tunnel into the other one.

The continua of these IFIGS are an *intrinsic* property of the semiconductors and they are the *fundamental* mechanism that determines both the barrier heights of Schottky contacts and the band offsets of semiconductor heterostructures. The IFIGS derive from the valence-band and conduction-band states of the semiconductor. The sign and the amount of the net charge in the IFIGS depend on the Fermi-level position relative to their branch point where their character changes from predominantly valence-band-like or donor-like to mostly conduction-band-like or acceptor-like. Hence, the IFIGS give rise to intrinsic interface dipoles. Both Schottky barrier heights and band offsets in heterostructures thus divide into a zero-charge-transfer term and a dipole contribution.

From a more chemical point of view, these interface dipoles are attributed to the partial ionic character of the covalent bonds between interface atoms. In generalizing Pauling's electronegativity concept [8], the difference of the electronegativities of the atoms involved in the interfacial bonds then describes the charge transfer at semiconductor interfaces. In combining the physical IFIGS and the chemical electronegativity concept, the dipole contributions of the Schottky barrier heights as well as the heterostructure-band offsets vary proportional to the difference of the electronegativities of the metal and the semiconductor and of the two semiconductors that are in contact, respectively.

Theoreticians appreciated Heine's IFIGS concept at once, but the experimentalists adopted it very slowly. One of the reasons was that the theoretical IFIGS lines marked upper limits of the barrier heights of *real* Schottky contacts only [9,10]. Schmitsdorf et al. [11] resolved this dilemma. They found a linear decrease of the effective barrier heights with increasing ideality factors of their Ag/n-Si(1 1 1) diodes. Such a behavior is observed with all Schottky contacts investigated so far. Schmitsdorf et al. attributed this correlation to patches of decreased barrier heights and lateral dimensions smaller than the depletion-layer width. Consequently, they extrapolated their plots of effective barrier heights vs the ideality factors to the ideality factor that is determined by the image force or Schottky effect [12] only and, in this way, obtained the barrier heights of laterally homogenous contacts.

The barrier heights of laterally uniform contacts may also be determined by applying ballistic electron emission microscopy (BEEM) and internal photoemission yield spectroscopy (IPEYS). The I/V, BEEM, and IPEYS data agree within the margins of experimental error. Mönch [13–15] plotted the barrier heights of laterally homogenous Si, GaN, GaP, GaAs, ZnSe, and 3C-SiC, 6H-SiC, and 4H-SiC Schottky contacts vs the difference of the metal and the semiconductor electronegativities. He found excellent agreement of the experimental data with the predictions of the IFIGS and electronegativity theory.

The IFIGS dipole term or, in other words, the difference between the metal and semiconductor electronegativities determines the dependence of the barrier heights of Schottky contacts with different metals on one and the same semiconductor. The electronegativities of the semiconductors are equal to within 10% since the elements that constitute the semiconductors are all placed in the middle of the Periodic Table of the Elements. Hence, the IFIGS dipole term of semiconductor heterostructures will be small and may be neglected [16]. The valence-band offsets of lattice-matched and non-polar as well as metamorphic heterostructures should thus equal the difference of the branch-point energies of the semiconductors in contact. The experimentally observed valence-band offsets of semiconductor heterostructures excellently confirm this prediction of the IFIGS-and-electronegativity theory [15].

This chapter is organized such that first a database of experimental barrier heights and valence-band offsets of I–III–VI$_2$ Schottky contacts and heterostructures, respectively, is compiled. Section 2.3 describes the IFIGS-and-electronegativity theory of the band-structure lineup at semiconductor interfaces. Section 2.4 is devoted to a comparison of experimental and theoretical data.

2.2 Experimental I–III–VI$_2$ Database

2.2.1 Barrier Heights of I–III–VI$_2$ Schottky Contacts

The barrier heights of Schottky contacts are generally determined from their current–voltage and capacitance–voltage characteristics (I/V, C/V) and by applying IPEYS and BEEM. No BEEM studies of I–III–VI$_2$ Schottky contacts have been published so far. Therefore, the evaluation of I/V, C/V and IPEYS characteristics will be outlined only briefly.[1]

I/V Characteristics

The current transport in real Schottky contacts occurs via thermionic emission provided the doping level of the semiconductor is not too high. The current–voltage characteristics may then be written as (see [15] for example):

$$I = AA_R^*T^2 \exp(-\varPhi_{Bn}^{eff}/k_BT)\exp(e_0V_c/nk_BT)[1 - \exp(-e_0V_c/k_BT)], \quad (2.1)$$

where A is the diode area, A_R^* is the effective Richardson constant of the semiconductor, and k_B, T, and e_0 are Boltzmann's constant, the temperature, and the electronic charge, respectively. The externally applied bias V_a divides up

---

[1] For a more detailed description of these techniques see [15].

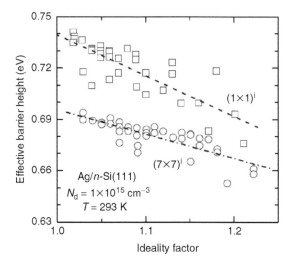

Fig. 2.1. Effective barrier heights vs ideality factors determined from I/V curves of Ag/n-Si(1 1 1)-$(7 \times 7)^i$ and Ag/n-Si(1 1 1)-$(1 \times 1)^i$ contacts at room temperature. The *dashed and dash-dotted lines* are linear least-squares fits to the data. From Schmitsdorf et al. [11]

into a voltage drop V_c across the depletion layer of the Schottky contact and an IR drop at the series resistance R_s of the diode, i.e., $V_c = V_a - IR_s$. For *ideal*, i.e., intimate, abrupt, defect-free, and above all, laterally homogenous Schottky contacts, the effective zero-bias barrier height Φ_{Bn}^{eff} equals the difference $\Phi_{Bn}^{hom} - \delta\Phi_{if}^0$ of the homogenous barrier height and the zero-bias image-force lowering. The ideality factor n describes the voltage dependence of the barrier height. For *real* diodes the ideality factors n are generally larger than the ideality factor n_{if}, which is determined by the image-force effect only.

The effective barrier heights and the ideality factors of real Schottky diodes fabricated under experimentally identical conditions differ from one specimen to another. However, the variations of both quantities are correlated. As an example, Fig. 2.1 displays effective barrier heights plotted vs the ideality factors of two sets of Ag/n-Si(1 1 1) contacts at room temperature. They differ in that the Si interface layers are either $(1 \times 1)^i$-unreconstructed or exhibit a $(7 \times 7)^i$ reconstruction. Both data sets reveal a pronounced correlation between the effective barrier heights and the ideality factors in that the effective barrier heights become smaller as the ideality factors increase. The dashed and dash-dotted lines are linear least-squares fits to the data. The dependence of the effective barrier heights on the ideality factors may thus be written as [11]

$$\Phi_{Bn}^{eff} = \Phi_{Bn}^{nif} - \varphi_p(n - n_{if}), \qquad (2.2)$$

where Φ_{Bn}^{nif} is the barrier height at the ideality factor n_{if}, which is determined by the image-force effect only. The diodes with $(1 \times 1)^i$-unreconstructed interfaces have a larger Φ_{Bn}^{nif} value than the contacts with $(7\times7)^i$-reconstructed interfaces.

Several conclusions may be immediately drawn from relation (2.2). First, the correlation between effective barrier heights and ideality factors demonstrates the existence of more than just one physical mechanism that determines the barrier heights of *real* Schottky contacts. Second, the extrapolation of Φ_{Bn}^{eff} vs n curves to n_{if}, the ideality factor controlled by the image-force effect only, leaves all effects out of consideration, which causes a larger bias dependence of the barrier height than the image-force effect itself. Third, the extrapolated barrier heights Φ_{Bn}^{nif} are equal to the zero-bias barrier height $\Phi_{Bn}^{0} = \Phi_{Bn}^{hom} - \delta\Phi_{if}^{0}$. The superscript hom indicates that the barrier heights are laterally uniform or homogenous.

The homogenous barrier heights obtained from extrapolations of Φ_{B}^{eff} vs n curves to n_{if}, the ideality factor controlled by the image-force effect only, are not necessarily the barrier heights of the corresponding *ideal* contacts. This is illustrated by the two data sets displayed in Fig. 2.1. The corresponding diodes differ in their interface structures, $(1 \times 1)^i$-unreconstructed and $(7 \times 7)^i$-reconstructed. Generally, structural rearrangements are connected with a redistribution of the valence charge. The bonds in perfectly ordered bulk silicon (the example considered here) are purely covalent and, therefore, reconstructions are accompanied by $\mathrm{Si}^{-\Delta q}$–$\mathrm{Si}^{+\Delta q}$ dipoles. In a simple point-charge model, reconstruction-displaced and then charged-silicon interface atoms may be treated in the same way as foreign atoms at interfaces: The electronegativities of the foreign and the semiconductor-substrate atoms generally differ so that they induce interfacial dipoles. Depending on their orientation, such extrinsic dipole layers increase or lower the barrier heights (see [15] for example).

Patches of reduced barrier height with lateral dimensions smaller than the depletion layer width, which are embedded in large areas of laterally homogenous barrier height is the only model known that explains a lowering of effective barrier heights with increasing ideality factors. In their phenomenological studies of such patchy Schottky contacts, Freeouf et al. [17, 18] found the potential distribution to show a saddle point *in front* of such nm-size patches of reduced barrier height. Figure 2.2 explains this behavior. For example, in front of circular patches, the barrier height right at the saddle point is lowered with respect to the laterally homogenous barrier height of the embedding area [19] by

$$\delta\Phi_{\pi}^{sad} = \gamma_{\pi}\left[(\Phi_{Bn}^{hom} - W_n - e_0V_c)k_BT/L_D^2\right]^{1/3}, \qquad (2.3)$$

where $W_n = W_{cb} - W_F$ and L_D are the energy distances from the Fermi level to the conduction-band edge in the bulk and the Debye length of the semiconductor, respectively. The saddle-point barrier height is determined by the patch parameter

$$\gamma_\pi = 3(\Delta_\pi R_\pi^2/4)^{1/3}, \qquad\qquad (2.4)$$

where R_π is the radius and Δ_π the barrier-height reduction of the patch.

Relation 2.3 shows that the saddle-point barrier height strongly depends on the voltage drop V_c across the depletion layer. Already Freeouf et al. [17,18] simulated the current transport in such patchy Schottky contacts and found a reduction of the effective barrier height and a correlated increase of the ideality factor as they reduced the lateral dimensions of the patches. Figure 2.3 displays some of their results, which they only tabulated. Unfortunately, Freeouf et al. missed to note that the barrier heights of the laterally homogenous contacts may be obtained from $\Phi_{\mathrm{Bn}}^{\mathrm{eff}}$ vs n plots by extrapolation to n_{if}. It remained for Schmitsdorf et al. [11] to draw this conclusion.

Figure 2.3 predicts that for larger ideality factors the $\Phi_{\mathrm{Bn}}^{\mathrm{eff}}$ vs n plots tend to level off. The experimental data of Ag/n-GaN(0 0 1) diodes, which are shown in Fig. 2.4 as an example, indeed prove the expected behavior. Obviously, the effective barrier heights and the ideality factors of real Schottky diodes are linearly correlated up to ideality factors of approximately 1.4 only.

Unfortunately, the I/V characteristics of very few I–III–VI$_2$ Schottky contacts have been studied and evaluated in detail. Table 2.1 shows the data available. Mostly, effective barrier heights and ideality factors of only one diode were communicated. The ideality factors are generally much larger than 1.01–1.05, the range of the ideality factor n_{if} is determined by the image-force effect only. This finding emphasizes the difficulties in the fabrication of I–III–VI$_2$ Schottky contacts. In Fig. 2.5, effective barrier heights of Al/p-CuInSe$_2$ and In/p-CuInSe$_2$ Schottky contacts are displayed as a function of the ideality factors. These data do not facilitate a reliable determination of the barrier heights of the corresponding laterally homogenous contacts. The dash-dotted

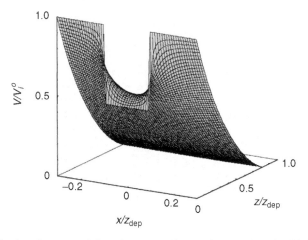

Fig. 2.2. Calculated potential distribution underneath a patch embedded in a region of larger interface band-bending or larger barrier height

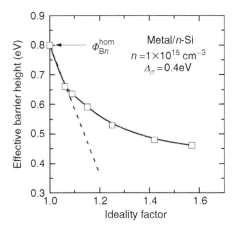

Fig. 2.3. Effective barrier height as a function of the ideality factor obtained from simulated current–voltage characteristics of laterally inhomogenous metal/n-Si contacts. The contacts possess one stripe with a reduced barrier height of 0.4 eV and lateral widths of 0.0313, 0.0625, 0.125, 0.25, 0.5, and 1 μm. The data in *open squares* are from Freeouf et al. [17] and the *dashed line* is a result of Tung's [19] analytical solution

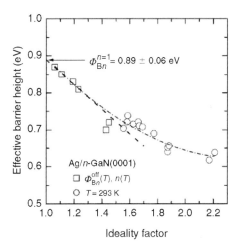

Fig. 2.4. Effective barrier heights vs ideality factors of Ag/n-GaN(0001) diodes. Data in *open squares* are of one diode at different temperatures between 150 and 400 K, from Sawada et al. [20]; data in *open circles* are of 13 different diodes at room temperature, from Kampen and Mönch [21]. The *dashed line* is a linear least-squares fit to the data points up to $n = 1.2$ while the *dash-dotted line* is only meant to guide the eye

Table 2.1. Barrier heights and ideality factors of I–III–VI$_2$ Schottky contacts, in eV

semiconductor	metal	Φ_{Bn}	Φ_{Bp}	n	method	reference
CuGaS$_2$	Ni		0.60	1.21	I/V	[22]
	Cu		0.90	1.5	I/V	[26]
			0.90		C/V	[26]
	Au		0.56	1.22	I/V	[22]
CuGaSe$_2$	Cu		0.60	3.0	I/V	[23]
CuInS$_2$	Au	0.85		1.25	C/V	[25]
CuInSe$_2$	Al		0.52	1.04	I/V	[23]
	Al		0.48	5	C/V	[31]
	Al		0.80		C/V	[34]
	Al		> 0.64	> 3.5	I/V	[29]
	Ni		0.55	1.04	I/V	[22]
	Au	0.36		1.74	C/V	[29]
	Au	0.54		1.3	I/V	[30]
	Au	0.47			C/V	[30]
	Au	0.50	0.60		SXPS	[42]
	Au	0.35		1.9	C/V	[32]
		0.83			C/V	[32]
		0.60			IPEYS	[32]
	In		1.02	2.3	C/V	[33]
	In		0.62	–	I/V	[35]
	In		> 0.65	> 1.9	I/V	[27]
	Mo		0.80		XPS	[42]
AgGaSe$_2$	Al	0.62		1.13	I/V	[22]
	Al	0.55		> 3.6	IPEYS	[38]
	Al	0.60		5	I/V	[24]
	Ni	0.49		1.6	I/V	[22]
	Ag		0.78	1.6	I/V	[24]
	Ag		1.04	2.2	C/V	[24]
	Ag		0.79		IPEYS	[24]
AgInSe$_2$	Ni	0.55		1.95	C/V	[40]
	Ni	0.60			IPEYS	[40]
AgInTe$_2$	Al		0.28	> 2.39	IPEYS	[39]

line is linear interpolation of the two In data points having the lowest ideality factors. The extrapolation of this line to $n \cong 1$ gives a numerical value of 0.65 eV. This figure is a rough estimate only and is certainly smaller than the true barrier height of laterally homogenous In/p-CuInSe$_2$ contacts. The Ag/n-GaN data plotted in Fig. 2.4 illustrates this conclusion. Kampen and Mönch [21] obtained diodes with ideality factor larger than 1.6 only and estimated the barrier height of the laterally homogenous contacts as 0.82 eV. The contacts fabricated by Sawada et al. [20] showed ideality factors as low as 1.06 and, as a consequence, the extrapolation of their data gave a larger value (0.89 eV) of the homogenous barrier height.

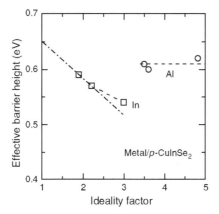

Fig. 2.5. Effective barrier heights vs ideality factors of Al/p-CuInSe$_2$ and In/p-CuInSe$_2$ diodes. The *dash-dotted line* is a liner least-squares fit to the first two data points while the *dashed lines* are only meant to guide the eye. Data from Matsushita et al. [27]

C/V Characteristics

The space charge and width of the depletion layer at a metal–semiconductor contact both vary as a function of the externally applied voltage. The depletion layer thus behaves like a parallel-plate capacitor. The differential capacitance per unit area of depletion layers at metal–semiconductor contacts is obtained as (see [15] for example)

$$C_{\text{dep}} = \partial Q_{\text{sc}}/\partial V_{\text{c}} = \{e_0^2 \varepsilon_{\text{b}} \varepsilon_0 N_{\text{d}}/[2e_0(|V_{\text{i0}}| - V_{\text{c}}) - k_{\text{B}}T]\}^{1/2}, \qquad (2.5)$$

where, ε_{b} and N_{d} are the static dielectric constant and the donor density of the semiconductor, respectively, $e_0|V_{\text{i0}}|$ is the zero-bias band bending at the interface, and ε_0 is the permittivity of vacuum. Hence, the inverse square $1/C_{\text{dep}}^2$ of the depletion-layer capacitance per unit area varies proportionally to the voltage drop across the barrier layer.

The current through Schottky contacts biased in the reverse direction is small, so that the IR drop due to the series resistance of the diode may be neglected. The voltage drop V_{c} across the depletion layer then equals the externally applied voltage V_{a}. Consequently, the extrapolated intercepts on the abscissa of $1/C^2$ vs V_{a} plots give the band bending $e_0|V_{\text{i0}}|$ at the interface. The extrapolation is equivalent to flat bands up to the interface, so that the intercept $V_{\text{a}}^{\text{int}}$ provides the flat-band barrier height

$$\Phi_{\text{Bn}}^{\text{fb}} = e_0|V_{\text{i0}}| + (W_{\text{cb}} - W_{\text{F}}) = e_0 V_{\text{a}}^{\text{int}} + k_{\text{B}}T + (W_{\text{cb}} - W_{\text{F}}) \qquad (2.6)$$

of the diode, where $W_{\text{cb}} - W_F$ is the energy distance from the Fermi level to the conduction-band minimum in the bulk.

The patches of lowered barrier height make up for a few percent of the total diode area only. Hence, the flat-band barrier height should be equal to the homogenous barrier height of the contact minus the zero-bias image-force lowering. The flat-band or C/V barrier heights of individual Schottky contacts are generally larger than their effective or I/V barrier heights. However, only the average flat-band barrier height of a sufficiently large number of identically prepared Schottky diodes provides a reliable value of their homogenous barrier height [28].

The reported flat-band or C/V barrier heights of I–III–VI$_2$ Schottky diodes are listed in Table 2.1.

Internal Photoemission Yield Spectroscopy

Metal–semiconductor contacts show a photoelectric response to optical radiation with photon energies far lower than the width of the bulk band gap. The primary process of this effect is the photoexcitation of hot electrons in the metal. Provided their energy is sufficiently large, they may surmount the interfacial barrier and reach the semiconductor. Schottky diodes thus have photothresholds that equal their barrier heights. This effect is called internal photoemission yield spectroscopy (IPEYS).

The internal photoemission yield $Y(\hbar\omega)$ is defined as the ratio of the photoemitted electron flux across the barrier into the semiconductor and the flux of the electrons excited in the metal by the photons of energy $\hbar\omega$. Cohen et al. [36] considered a very simple model and derived the spectral dependence of the internal photoemission yield (see [15] for example) as

$$Y(\hbar\omega) \propto \frac{(\hbar\omega - \hbar\omega_Y)^2}{\hbar\omega}, \qquad (2.7)$$

where $\hbar\omega_Y$ is the threshold energy. Fowler [37] calculated the temperature dependence of the photoemission yield for clean metal surfaces. His zero-temperature result $Y_m \propto (\hbar\omega - \hbar\omega_Y)^2$ differs from relation 2.7 in that the photon energy $\hbar\omega$ in the denominator is missing. Nevertheless, Fowler's function was mostly used in determining the IPEYS thresholds of Schottky contacts. The patches of lowered barrier height generally make up for some percent of the total diode area only. Hence, the IPEYS barrier height $\Phi_Y = \hbar\omega_Y$ plus the zero-bias image-force lowering should be equal to the homogenous barrier height of the contact. The reported IPEYS barrier heights of I–III–VI$_2$ Schottky diodes are listed in Table 2.1.

2.2.2 Valence-Band Offsets of I–III–VI$_2$ Heterostructures

Semiconductors generally grow layer by layer, at least initially. Hence, core-level photoemission spectroscopy is one of the most reliable tools for determining the band-structure lineup at semiconductor heterostructures. The valence-band offset may be obtained from the energy positions of core-level lines in

Table 2.2. Valence-band offsets of I–III–VI$_2$ heterostructures, in eV

heterostructure		ETB	LAPW	experiment	reference
layer 1	layer 2				
CuInS$_2$	Si	1.11		0.08	[54]
	ZnS	0.58	(095)[a]	2.30	[49]
	CdS	0.46	(0.79)[a]	0.60	[47]
	CuI			0.10	[55]
CuGaSe$_2$	ZnSe	0.55	(0.66)[a]	0.60	[52]
	CdS	1.00	(1.03)[a]	0.90	[50]
CuInSe$_2$	SnO$_2$			2.60	[48]
	ZnS	1.30	(1.23)[a]	1.40	[49]
	ZnSe	0.73	0.70[b]	0.70	[44]
				0.50	[49]
	CdS	1.18	1.07[b]	0.80	[43, 45]
				1.06	[46]
				1.20	[51]
				0.80	[53]
	CuInS$_2$	0.72	0.28[a]	0.87	[49]
	CuInTe$_2$	0.20	0.50[a]	0.85	[49]

[a]Ref. [57]
[b]Ref. [56]
The experimental values were determined by using XPS. The ETB values are
from Table 2.3. The LAPW values in brackets assumed the transitivity rule to
hold.

X-ray photoelectron spectra (XPS) recorded with bulk samples of the semiconductors in contact and with the interface itself [41]. Since the escape depths of
the photoelectrons are in the order of 2 nm only, one of the two semiconductors
has to be sufficiently thin. This condition is easily met when heterostructures
are grown by molecular beam epitaxy (MBE) and XPS spectra are recorded
during growth interrupts. The valence-band discontinuity results as

$$\Delta W_{\rm v} = W_{\rm vir} - W_{\rm vil} = W_{\rm ir}(n_r l_r) - W_{\rm il}(n_l l_l) + [W_{\rm vbr} - W_{\rm br}(n_r l_r)]$$
$$-[W_{\rm vbl} - W_{\rm bl}(n_l l_l)], \tag{2.8}$$

where $n_r l_r$ and $n_l l_l$ denote the core levels of the semiconductors on the right-
hand side (r) and the left-hand side (l) of the interface, respectively. Subscripts
i and b characterize interface and bulk properties, respectively. The energy difference $W_{\rm ir}(n_r l_r) - W_{\rm il}(n_l l_l)$ between the core levels of the two semiconductors
is determined from EDCs of photoelectrons recorded during MBE growth of
the heterostructure. The energy positions $W_{\rm vbr} - W_{\rm br}(n_r l_r)$ and $W_{\rm vbl} - W_{\rm bl}(n_l l_l)$
of the core levels relative to the valence-band maxima in each of the two

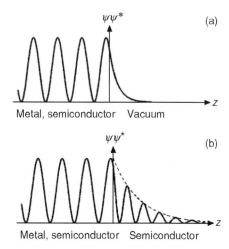

Fig. 2.6. Wave-function tails at (**a**) surfaces and (**b**) at semiconductor interfaces in energy regions where occupied states overlap a band gap

semiconductors are evaluated separately. Table 2.2 shows valence-band offsets of I–III–VI$_2$ heterostructures that were obtained by core-level XPS.

The XPS technique may also be applied to metal–semiconductor contacts. Table 2.1 also shows homogenous barrier heights of I–III–VI$_2$ Schottky contacts that were determined by using X-ray and soft-X-ray photoemission spectroscopy.

2.3 IFIGS-and-Electronegativity Theory

Because of the quantum-mechanical tunnel effect, the wave functions of bulk electrons decay exponentially into vacuum at surfaces or, more generally, at solid–vacuum interfaces. A similar behavior occurs at interfaces between two solids. In energy regions of Schottky contacts and semiconductor heterostructures where occupied band states overlap a band gap, the wave functions of these electrons tails across the interface. The only difference with respect to solid–vacuum interfaces is that the wave-function tails oscillate at solid–solid interfaces. Figure 2.6 schematically explains the tailing effects at surfaces and semiconductor interfaces. For the band-structure lineup at semiconductor interfaces, only the tailing states within the gap between the top valence and the lowest conduction band are of importance since the energy position of the Fermi level determines their charging state. These wave-function tails or IFIGS derive from the continuum of the virtual gap states (ViGS) of the complex semiconductor band structure. Hence, the IFIGS are an intrinsic property of the semiconductor.

The IFIGS are made up of valence-band and conduction-band states of the semiconductor. Their net charge depends on the energy position of the Fermi

Table 2.3. Optical dielectric constants, energies of the valence-electron plasmons, widths of the dielectric band gap, and branch-point energies of selected semiconductors and I–III–VI$_2$ chalcopyrites

semiconductor	ε_∞	$\hbar\omega_p$ [eV]	W_{dg} [eV]	$[W_{bp} - W_v(\Gamma)]_{\mathrm{ETB}}$ [eV]
Si	11.90	16.90	5.04	0.36[a]
ZnS	5.14	16.69	8.12	2.05
ZnSe	5.70	15.57	7.06	1.48
CdS	5.27	14.97	7.06	1.93
CuAlS$_2$	–	17.10	–	–
CuGaS$_2$	6.15	16.93	7.46	1.43
CuInS$_2$	6.3[b]	16.16	7.02	1.47
CuAlSe$_2$	6.3[b]	15.78	6.85	1.25
CuGaSe$_2$	7.3[b]	15.78	6.29	0.93
CuInSe$_2$	9.00	15.11	5.34	0.75
CuAlTe$_2$	–	–	–	–
CuGaTe$_2$	8.0[b]	14.25	5.39	0.61
CuInTe$_2$	9.20	13.68	4.78	0.55
AgAlSe$_2$	–	–	–	–
AgGaSe$_2$	6.80	14.35	5.96	1.09
AgInSe$_2$	7.20	13.94	5.60	1.11

[a]Ref. [61]
[b]$\varepsilon_\infty = n^2$

level relative to their branch point where their character changes from predominantly donor-like or valence-band-like to mostly acceptor-like or conduction-band-like. The band-structure lineup at semiconductor interfaces is thus described by a zero-charge-transfer term and an electric-dipole contribution.

The charge transfer at semiconductor interfaces may be easily estimated by applying Pauling's description of the partial ionic character of covalent bonds by the difference of the electronegativities of the atoms involved. The combination of the physical IFIGS and the chemical electronegativity concept yields the barrier heights of ideal p-type Schottky contacts to vary as

$$\Phi_{\mathrm{Bp}} = \Phi_{\mathrm{bp}}^{\mathrm{p}} - S_X(X_{\mathrm{m}} - X_{\mathrm{s}}) \tag{2.9}$$

and the valence-band offsets of ideal semiconductor heterostructures to vary as

$$\Delta W_{\mathrm{v}} = \Phi_{\mathrm{bpr}}^{\mathrm{p}} - \Phi_{\mathrm{bpl}}^{\mathrm{p}} + D_X(X_{\mathrm{sr}} - X_{\mathrm{sl}}), \tag{2.10}$$

where $\Phi_{\mathrm{bp}}^{\mathrm{p}} = W_{\mathrm{bp}} - W_{\mathrm{v}}(\Gamma)$ is the energy distance from the valence-band maximum to the branch point of the IFIGS or the p-type branch-point energy. It has the physical meaning of a zero-charge-transfer barrier height. X_{m} and X_{s} are the electronegativities of metals and semiconductors, respectively. The subscripts r and l stand for the right-hand side and the left-hand side, respectively, of heterostructures. The slope parameters S_X and D_X are explained at the end of this section.

The IFIGS derive from the virtual gap states of the complex band structure of the semiconductor. Their branch point is an average property of the semiconductor. Bardeleschi's concept [58] of mean-value k-points in the Brillouin zone avoids the extensive computations required for calculations of branch points. At the mean-value k-point, the separation between the valence and the conduction bands indeed equals the average or dielectric band gap [59]

$$W_{dg} = \hbar\omega_p/\sqrt{\varepsilon_\infty - 1}, \qquad (2.11)$$

where $\hbar\omega_p$ is the plasmon energy of the bulk valence electrons and ε_∞ is the optical dielectric constant [60]. Mönch [59] plotted the theoretical branch-point energies $W_{bp} - W_v(\underline{k}_{mv}) = \Phi^p_{bp} + [W_v(\Gamma) - W_v(\underline{k}_{mv})]_{ETB}$ at the mean-value k-point $\underline{k}_{mv}$ vs the width of the dielectric band gaps of 15 different elemental and compound semiconductors. He took the branch-point energies $\Phi^p_{bp} = W_{bp} - W_v(\Gamma)$ that had been computed by Tersoff [61] for Si, Ge, and 13 of the II–VI and III–V compound semiconductors and calculated the energy dispersions $[W_v\Gamma) - W_v(k_{mv})]_{ETB}$ of the top valence band in the empirical tight-binding approximation. A linear least-squares fit to the data yielded [59]

$$\Phi^p_{bp} = 0.449 \cdot W_{dg} - [W_v(\Gamma) - W_v(\underline{k}_{mv})]_{ETB}. \qquad (2.12)$$

The IFIGS branch points are thus slightly below the middle of the band gap at the mean-value k-point. The same result is well known for the ViGS of one-dimensional linear chains.

A simple phenomenological model of Schottky contacts with a continuum of interface states of constant density of states D_{is} across the semiconductor band gap yields the slope parameter [62, 63]

$$S_X = A_X/[1 + (e_0^2/\varepsilon_i\varepsilon_0)D_{is}\delta_{is}], \qquad (2.13)$$

where ε_i is an interface dielectric constant. The parameter A_X depends on the electronegativity scale chosen and amounts to $0.86\,\mathrm{eV/Miedema}$ unit and $1.79\,\mathrm{eV/Pauling}$ unit. The extension δ_{is} of the interface states is approximated by their charge decay length $1/2q_{is}$. For one-dimensional linear chains, both the density of states and the charge decay lengths of their ViGS vary inversely proportional to the width of their band gaps. In view of relations 2.11 and 2.13, Mönch [63] plotted theoretical values $(e_0^2/\varepsilon_0)D^{mi}_{gs}/2q^{mi}_{gs}$ of metal-induced gap states (MIGS, as the IFIGS in Schottky contacts are traditionally called) vs the optical susceptibility $\varepsilon_\infty - 1$. A least-squares fit to the data points yielded a quadratic dependence, and relation 2.13 was rewritten as

$$A_X/S_X - 1 = 0.1 \cdot (\varepsilon_\infty - 1)^2, \qquad (2.14)$$

where the reasonable assumption $\varepsilon_i \approx 3$ was made.

To a first approximation, the slope parameter D_X of heterostructures may be equated with the slope parameter S_X of Schottky contacts since the IFIGS determine the intrinsic dipole contributions to both the valence-band offsets and the barrier heights. Furthermore, the elements forming semiconductors are all found in the center columns of the Periodic Table of the Elements. Therefore, their electronegativities are almost equal so that the dipole term $D_X(X_{sr} - X_{sl})$ in Eq. 2.10 will be small and may be neglected. Consequently, Eq. 2.10 reduces to

$$\Delta W_v \cong \Phi_{bpr}^p - \Phi_{bpl}^p \tag{2.15}$$

for practical purposes.

The simple calculation of branch-point energies according to relation 2.12 requires the widths W_{dg} of the dielectric band gaps and the valence-band dispersions $[W_v(\Gamma) - W_v(\underline{k}_{mv})]_{ETB}$ of the semiconductors. In analogy to ternary and quaternary III–V alloys, Mönch [15] calculated the ETB valence-band energies of the I–III–VI₂ chalcopyrites by assuming virtual I–III cations. The last column of Table 2.3 displays his results.

2.4 Comparison of Experiment and Theory

2.4.1 IFIGS-and-Electronegativity Theory

I–III–VI₂ Schottky Contacts

The experimental barrier heights of intimate, abrupt, clean and laterally homogenous Schottky contacts on p-CuInSe₂ are plotted vs the difference of the Miedema electronegativities of the metals and CuInSe₂ in Fig. 2.7. Miedema's

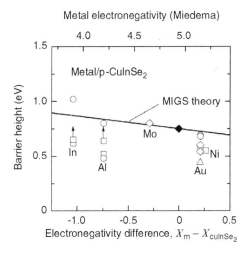

Fig. 2.7. Barrier heights of p-CuInSe₂ Schottky contacts vs the difference of the Miedema electronegativities of the metal and CuInSe₂. *Open squares I/V data, open circles C/V data, open diamonds XPS data, and open triangles IPEYS data* are from Table 2.1. The *solid MIGS line* is drawn with the p-type zero-charge-transfer barrier height 0.75 eV and the slope parameter 0.116 eV/Miedema unit

electronegativities are preferred since they were derived from properties of metal alloys and intermetallic compounds, while Pauling considered covalent bonds in small molecules. The electronegativity of a compound is taken as the geometric mean of the electronegativities of its constituent atoms. The Miedema electronegativity of selenium is estimated as 5.8 Miedema units from the corresponding Pauling value, 2.55 Pauling units, and the correlation

$$X_{\mathrm{Mied}} = 1.93 X_{\mathrm{Paul}} + 0.87 \qquad (2.16)$$

between the two scales. The MIGS line results from the p-type zero-charge-transfer barrier height $\Phi_{\mathrm{bp}}^{\mathrm{p}} = W_{\mathrm{bp}} - W_{\mathrm{v}}(\Gamma) = 0.75$ eV and the slope parameter $S_{\mathrm{X}} = 0.116\,\mathrm{eV/Miedema}$ unit of $CuInSe_2$.

The MIGS and electronegativity theory describes quite well both the values and the trend of the available barrier heights of p-$CuInSe_2$ Schottky contacts. The experimental data are either on or below the MIGS line. The latter deviations toward smaller barrier heights may be again explained by patches of lowered barrier height and lateral dimensions smaller than the depletion-layer width. The arrows are put on the I/V data points of the In-$CuInSe_2$ and Al/p-$CuInSe_2$ contacts that are plotted in Fig. 2.5. They indicate that the corresponding laterally homogenous contacts will have larger barrier heights, which then are in closer agreement with the predictions of the MIGS-and-electronegativity theory.

I–III–VI$_2$ Heterostructures

The experimental valence-band offsets of I–III–VI$_2$ heterostructures, which are compiled in Table 2.2, are plotted vs the corresponding differences of the ETB branch-point energies of the IFIGS in Fig. 2.8. The dashed line indicates that the IFIGS theory explains the experimental data of the I–III–VI$_2$ heterostructures. The $ZnS/CuInS_2$ and $CuInTe_2/CuInSe_2$ results are not considered since they strongly deviate from the general trend.

Morkel et al. [53] studied technical devices and reported one of the two data points for $CdS/CuInSe_2$ that deviate by approximately 0.3 eV to lower values. Heske et al. [64] additionally investigated such heterostructures using a combination of X-ray emission and X-ray photoemission spectroscopy. Their studies revealed an anion intermixing at the interface of these cells. This effect reduces the difference of the corresponding branch-point energies and brings the experimental valence-band offset in closer agreement with the prediction of the IFIGS theory. This is indicated by the vertical arrow.

Hunger et al. [54] prepared $Si(1\,1\,1)/CuInS_2$ heterostructures and obtained an *effective* valence-band offset of 0.08 ± 0.15 eV. They argued that this small value is caused by an interfacial layer of $S^{-\Delta q}-Si^{+\Delta q}$ dipoles that are oriented such as to reduce the *true* valence-band offset. This behavior is similar to what had been reported for ZnS/Si interfaces. Maierhofer et al. [65] investigated $ZnS/Si(1\,1\,1)$ heterostructures. They determined a valence-band offset

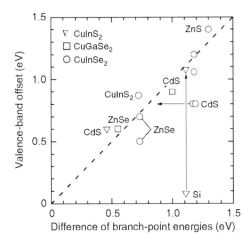

Fig. 2.8. Experimental valence-band offsets of I–III–VI₂ heterostructures vs the difference of the IFIGS branch-point energies. Data is from Tables 2.2 and 2.3 (last column), respectively. The dashed line indicates complete agreement between experimental and theoretical data

of 0.7 eV. Brar et al. [66] on the other hand, studied ZnS/Si(0 0 1) heterostructures and from their data a valence-band offset of 1.7 ± 0.04 eV resulted. The difference of approximately 1 eV between both values is easily explained by an $S^{-\Delta q}-Si^{+\Delta q}$ dipole layer that is present at the $(1\,1\,1)$-oriented interface but missing at the $(0\,0\,1)$-oriented interface.

The chemical bonds between interlayer and substrate atoms may be described as "interface molecules" [67]. Pauling [8] correlated the amount of ionic character Δq_1 of single bonds in diatomic molecules A–B with the difference X_A-X_B of the atomic electronegativities of the constituent atoms. In a simple point-charge model, the respective dipole moment then amounts to

$$p_{eid} = e_0 \Delta q_1 d_{A-B} = e_0 [0.16|X_A - X_B| + 0.035(X_A - X_B)^2](r_{cov}^A + r_{cov}^B), \quad (2.17)$$

where the bond length d_{A-B} is taken as the sum of the covalent radii of the constituent atoms. Provided the interfacial dipoles are oriented perpendicular to the interface, then the voltage drop across the extrinsic interfacial dipole layer equals the variation of the valence-band offset, i.e.,

$$\delta(\Delta W_v) = \pm \frac{e_0}{\varepsilon_i \varepsilon_0} p_{eid} N_{eid}, \quad (2.18)$$

where ε_i and N_{eid} are the dielectric constant and the number of extrinsic dipoles per unit area at the interface, respectively. For a monolayer of interfacial sulfur atoms covalently bound to silicon atoms, relation 2.18 yields $\delta(\Delta W_v) = 3.9/\varepsilon_i$ eV. With the reasonable assumption of $\varepsilon_i \approx 3$ for compound semiconductors [15], one then obtains a reduction of the valence-band offset by approximately 1.3 eV. This value is close to the experimentally observed difference of 1 eV between the valence-band offsets at ZnS/Si(1 1 1) and ZnS/Si(0 0 1) interfaces. Hence, an interfacial $S^{-\Delta q}-Si^{+\Delta q}$ dipole layer might lower the valence-band offset of Si(1 1 1)/CuInS$_2$ heterostructures by 1 eV to the experimentally observed value of 0.08 ± 0.15 eV. This then brings the estimated value, 1.08 ± 0.15 eV, in close agreement with the valence-band offset of 1.11 eV predicted for *intimate* Si(1 1 1)/CuInS$_2$ heterostructures by the IFIGS-and-electronegativity theory.

Langer–Heinrich Rule

In a heuristic approach, Langer and Heinrich [68] assumed the existence of a *common bulk reference level* for the energy levels of the 3d transition-metal impurities in isovalent semiconductors. Consequently, they concluded that a valence-band-edge discontinuity in heterojunctions is then given by just the difference in the energy-level positions of a transition-metal impurity in the two compounds forming the heterojunction. One of such examples is the Co impurity in CuGa$_{1-x}$Al$_x$Se$_2$ chalcopyrite alloys, which were investigated by Jin et al. [69] applying optical absorption spectroscopy. Their experimental data are displayed in Fig. 2.9. Evidently, their absorption thresholds are excellently described by the difference of the IFIGS branch-point energies of the CuGa$_{1-x}$Al$_x$Se$_2$ alloys and CuGaSe$_2$ [70]. The value for the CuGa$_{05}$Al$_{0.5}$Se$_2$ alloy was calculated assuming virtual CuGa$_{05}$Al$_{0.5}$ cations. However, it has to be mentioned that the experimentally determined threshold energies are assigned to electron transitions from the valence band into the Co impurity

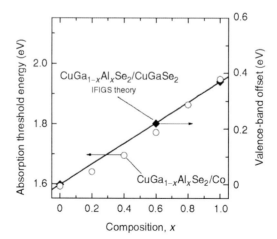

Fig. 2.9. Absorption threshold energies of deep Co impurities in CuGa$_{1-x}$Al$_x$Se$_2$ alloys and IFIGS valence-band offsets. Co impurity data from Jin et al. [69]; IFIGS data from Table 2.3 and from Mönch [70]

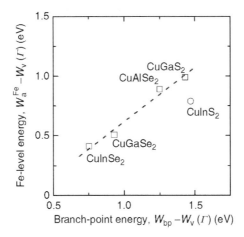

Fig. 2.10. Energy position of Fe acceptor levels above the valence-band maximum in I–III–VI$_2$ chalcopyrites as a function of the ETB branch-point energies. Acceptor levels are from [73–75] and ETB branch-point energies are from Table 2.3

levels [70] rather than from the Co levels into the conduction band as inferred earlier by Chang et al. [71].

Another example is Fe impurities in various I–III–VI$_2$ chalcopyrites [72]. Figure 2.10 displays the energy positions of the Fe acceptor levels above the valence-band maxima of the I–III–VI$_2$ compounds [73–76] as a function of the ETB branch-point energies displayed in the last column of Table 2.3. The dashed line is the linear least-squares fit

$$W_a^{Fe} - W_v(\Gamma) = -(0.30 \pm 0.10) + (0.91 \pm 0.09)[W_{bp} - W_v(\Gamma)]_{ETB} \text{ [eV]}$$

to the four *open square* data points. The slope parameter is close to unity, the value predicted by the Langer–Heinrich rule, but the *open circle* data point for Fe in CuInS$_2$ strongly deviates from the trend. Obviously, the Langer–Heinrich rule is not universally applicable but it holds in favorable cases (see also [15]):

ZnO Heterostructures

The most efficient Cu(In,Ga)Se$_2$ solar cells contain ZnO window layers. Hence, the valence-band offset between this window layer and either the CdS buffer layer or the Cu(In,Ga)Se$_2$ absorber layer is of interest. The available experimental data are compiled in Table 2.4, which also lists the valence-band offsets determined with some other semiconductors. Unfortunately, the ETB branch-point energy of ZnO is difficult to calculate. Considering relation 2.10, the experimental valence-band offsets of ZnO heterostructures are hence plotted vs the ETB branch-point energies of the corresponding second semiconductors of the heterostructures in Fig. 2.11 [86]. The valence-band offsets of the

28   W. Mönch

Table 2.4. Experimental valence-band offsets of ZnO heterostructures and branch-point energies of the corresponding semiconductors, in eV

semiconductor	$[W_{bp} - W_v(\Gamma)]_{ETB}$	ΔW_v	reference
Ge	0.18[a]	2.70	[78]
GaN	2.37[b]	0.80	[80]
ZnSe	1.48[b]	0.90	[79]
CdS	1.93[b]	1.20	[77]
CdS	1.93[b]	0.96	[85]
CuInSe$_2$	0.75[b]	2.20	[83]
CuGa$_{0.2}$In$_{0.8}$Se$_2$	0.79	2.30	[81]
CuGa$_{0.38}$In$_{0.62}$Se$_2$	0.82[b]	2.30	[82]
CuIn(S$_{0.75}$Se$_{0.25}$)$_2$	1.29[b]	1.80	[84], Heske, C. (private communication)

[a]Ref. [61].
[b]Ref. [15].

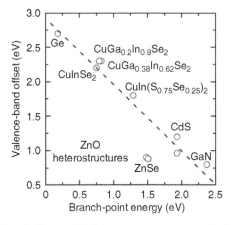

Fig. 2.11. Valence-band offsets of ZnO heterostructures vs the branch-point energies of the corresponding second semiconductor. The *dashed line* is a linear least-squares fit to the data. Data are from Table 2.4. From Mönch [86]

I–III–VI$_2$/ZnO heterostructures, Cu(In,Ga)Se$_2$/ZnO and CuIn(S$_{0.75}$Se$_{0.25}$)$_2$, excellently fit in the trend of the data observed with the other ZnO heterostructures. The dashed line is the linear least-squares fit [86] to the data. Within the margins of experimental error, the slope parameter has the value 1. The p-type branch-point energy of the ZnO results as 3.04 ± 0.21 eV.

$$\Delta W_v = \varphi_{vbo}[\Phi_{bp}^p(ZnO) - \Phi_{bp}^p(sem)]$$
$$= (0.94 \pm 0.15)[(3.04 \pm 0.21) - \Phi_{bp}^p(sem)] \text{ [eV]} \quad (2.19)$$

2.4.2 Linearized Augmented Plane Wave Method

In their calculations of valence-band offsets of I–III–VI$_2$ and II–VI heterostructures, Wei and Zunger [56,57] followed the same procedure used in core-level XPS, which is briefly described in Sect. 2.2. They calculated the separation of the corresponding core levels to the valence-band maxima in the pure semiconductors and the energy difference between corresponding core levels at the interface. They used the local-density-functional formalism as implemented by the general potential, relativistic all electron, linearized augmented plane wave (LAPW) method. Column 4 of Table 2.2 displays some of the results reported by Wei and Zunger [56,57]. The valence-band offsets predicted by the simple ETB and the first-principles LAPW approach are almost identical for five of the nine heterostructures considered in Table 2.2. These theoretical values also agree with the experimental data to within the margins of experimental error. The ETB and LAPW values differ for ZnS/CuInS$_2$, CdS/CuInS$_2$, CuInS$_2$/CuInSe$_2$, and CuInTe$_2$/CuInSe$_2$ interfaces. For the CdS/CuInS$_2$ and CuInS$_2$/CuInSe$_2$ heterostructures the ETB rather than the LAPW predictions are closer to the experimental values. The experimental valence-band offsets reported for the other two examples, ZnS/CuInS$_2$ and CuInTe$_2$/CuInSe$_2$, differ considerably from both theoretical predictions. Additional experimental studies are desirable.

2.5 Conclusions

In energy regions where occupied band states on one side overlap a band gap on the other side of a semiconductor interface, the quantum-mechanical tunnel effect causes the wave-functions of the conduction-band or valence-band electrons to tail across the interface. These intrinsic IFIGS determine the band lineup at ideal semiconductor interfaces and, as to be expected, the I–III–VI$_2$ chalcopyrites behave like all other semiconductors in that the IFIGS-and-electronegativity theory quantitatively describes both the barrier heights of I–III–VI$_2$ Schottky contacts and the valence-band offsets of I–III–VI$_2$ heterostructures. An essential for the comparison of experimental data and theoretical predictions is that experimental data are available for ideal interfaces. In the past, many of such tests failed since real interfaces, and this is especially true for Schottky contacts, contain defects of various origins. Unfortunately, the database for laterally homogenous I–III–VI$_2$ Schottky contacts is very limited, and additional experimental studies are very desirable. However, well-prepared and well-characterized interfaces and carefully chosen experimental methods combined with an elaborate analysis of the measured data make it possible to find agreement between experimental data obtained with *real* samples and results of calculations assuming ideal interfaces, even for such delicate semiconductors as the I–III–VI$_2$ chalcopyrites.

References

1. Braun, F.: Über die Stromleitung durch Schwefelmetalle. Pogg. Ann. Phys. Chem. 153, 556–563 (1874)
2. Schottky, W.: Halbleitertheorie der Sperrschicht. Naturwissenschaften 26, 843–843 (1938)
3. Mott, N.F.: Note on the contact between a metal and an insulator or semi-conductor. Proc. Camb. Philos. Soc. 34, 568–572 (1938)
4. Schottky, W.: Abweichungen vom Ohmschen Gesetz in Halbleitern. Phys. Zeitschr. 41, 570–573 (1940)
5. Bardeen, J.: Surface states and rectification at a metal semi-conductor contact. Phys. Rev. 71, 717–727 (1947)
6. Heine, V.: Theory of surface states. Phys. Rev. A 138, 1689–1696 (1965)
7. Tejedor, C., Flores, F.: A simple approach to heterojunctions. J. Phys. C 11, L19–L23 (1978)
8. Pauling, L.N.: The Nature of the Chemical Bond. Cornell University Press, Ithaca, NY (1939)
9. Mönch, W.: Role of virtual gap states and defects in metal-semiconductor contacts. Phys. Rev. Lett. 58, 1260–1263 (1987)
10. Mönch, W.: Chemical trends in Schottky barriers: Charge transfer into adsorbate-induced gap states and defects. Phys. Rev. B 37, 7129–7132 (1988)
11. Schmitsdorf, R., Kampen, T.U., Mönch, W.: Correlation between barrier height and interface structure of Ag/Si(111) Schottky diodes. Surf. Sci. 324, 249–256 (1995)
12. Schottky, W.: Über den Einfluß von Strukturwirkungen, besonders der Thom-sonschen Bildkraft, auf die Elektronenemission der Metalle,. Phys. Zeitschr. 15, 872–878 (1914)
13. Mönch, W.: Barrier heights of real Schottky contacts explained by metal-induced gap states and lateral inhomogenities. J. Vac. Sci. Technol. B 17, 1867–1876 (1999)
14. Mönch, W.: Semiconductor Surfaces and Interfaces, 3rd edn. Springer, Berlin Heidelberg New York (2001)
15. Mönch, W.: Electronic Properties of Semiconductor Interfaces. Springer, Berlin Heidelberg New York (2004)
16. Mönch, W.: In: Grosse, P. (ed.) Festkörperprobleme (Adv. Solid State Physics), vol. 26, p. 67. Vieweg, Braunschweig (1986)
17. Freeouf, J.L., Jackson, T.N., Laux, S.E., Woodall, J.M.: Effective barrier heights of mixed phase contacts: Size effects. Appl. Phys. Lett. 40, 634–636 (1982)
18. Freeouf, J.L., Jackson, T.N., Laux, S.E., Woodall, J.M.: Size dependence of 'effective' barrier heights of mixed-phase contacts. J. Vac. Sci. Technol. B 21, 570–573 (1982)
19. Tung, R.T.: Electron transport at metal-semiconductor interfaces: General theory. Phys. Rev. B 45, 13509–13523 (1992)
20. Sawada, T., Ito, Y., Imai, K., Suzuki, K., Tomozawa, H., Sakai, S.: Electrical properties of metal/GaN and SiO2/GaN interfaces and effects of thermal annealing. Appl. Surf. Sci. 159/160, 449–455 (2000)
21. Kampen, T.U., Mönch, W.: Barrier heights of GaN Schottky contacts. Appl. Surf. Sci. 117, 388–393 (1997)
22. Robinson, P., Wilson, J.I.B.: Schottky diodes on I-III-Se$_2$ compounds. Inst. Phys. Conf. Ser. 35, 229–236 (1977)

23. Stankiewicz, J., Giriat, W.: Photovoltaic effect in Cu/CuGaSe$_2$ Schottky barriers. Appl. Phys. Lett. 35, 70–71 (1979)

24. Murthy, Y.S., Hussain, O.M., Uthanna, S., Naidu, B.S., Reddy, P.J.: Electrical and photovoltaic properties of Ag/p-AgGaSe$_2$ polycrystalline thin film Schottky barrier diodes. Phys. Lett. 152, 311–313 (1991)

25. Kościelniak-Mucha, B., Opanowicz, A.: Electrical characteristics of Au/ n-CuInS$_2$ Schottky diodes. Phys. stat. sol. (a) 130, K55–K58 (1992)

26. Sugijama, M., Nakai, R., Nakanishi, H., Chichibu, S.F.: Fermi-level pinning at the metal/p-type CuGaS$_2$ interfaces. J. Appl. Phys. 92, 7317–7319 (2002)

27. Matsushita, H., Tojo, Y., Takizawa, T.: Schottky properties of CuInSe$_2$ single crystals grown by the horizontal Bridgman method with controlling Se vapor pressure. J. Phys. Chem Solids 64, 1825–1829 (2003)

28. Schmitsdorf, R., Mönch, W.: Influence of the interface structure on the barrier height of homogeneous Pb/n-Si(111) Schottky contacts. Eur. Phys. J. B 7, 457–466 (1999)

29. Parkes, J., Tomlinson, R.D., Hampshire, M.J.: Electrical properties of CuInSe$_2$ single crystals. Solid-State Electron. 16, 773–777 (1973)

30. Prasad, J.J.B., Rao, D.K., Sridevi, D., Majhi, J., Reddy, K.V., Sobhanadri, J.: Characteristics of n-CuInSe$_2$/Au Schottky diodes. Solid-State Electron. 28, 1251–1254 (1985)

31. Rao, D.K., Prasad, J.J.B., Sridevi, D., Reddy, K.V., Sobhanadri, J.: Properties of p-CuInSe$_2$/Al Schottky devices. Phys. stat. sol. (a) 94, K153–K158 (1986)

32. Opanowicz, A., Kościelniak-Mucha, B., Boroś, G.: Electrical and photovoltaic properties of Au/n-CuInSe$_2$ Schottky barrier diodes. Phys. stat. sol. (a) 106, K197–K201 (1988)

33. Opanowicz, A., Kościelniak-Mucha, B.: Photocapacitance effect in In/p-CuInSe$_2$ Schottky-type junction. Acta Phys. Pol. (A) 75, 289–292 (1989)

34. Chan, C.L., Shih, I.: Al/p-CuInSe$_2$ metal-semiconductor contacts. J. Appl. Phys. 68, 156–160 (1990)

35. Kościelniak-Mucha, B., Opanowicz, A.: Hole tunneling through the In/p-CuInSe$_2$ Schottky barrier. Phys. stat. sol. (a) 141, K67–K70 (1994)

36. Cohen, J., Vilms, J., Archer, R.J.: Hewlett-Packard R&D Report AFCRL-69-0287 (1969)

37. Fowler, R.H.: The analysis of photoelectric sensitivity curves for clean metals at various temperatures. Phys. Rev. 38, 45–56 (1931)

38. Patel, S.M., Kapale, V.G.: AgGaSe$_2$/Al thin film Schottky barrier diodes. Phys. stat. sol. (a) 101, K87–K92 (1987)

39. Patel, B.H.: AgInTe$_2$-Al Schottky diode. Phys. stat. sol. (a) 124, K155–K159 (1991)

40. Ramesh, P.P., Uthanna, S., Naidu, B.S., Reddy, P.J.: Characteristics of Ni/nAgInSe$_2$ polycrystalline thin film Schottky barrier diodes. Thin Solid Films 292, 290–292 (1997)

41. Grant, R.W., Waldrop, J.R., Kraut, E.A.: Observation of the orientation dependence of interface dipole energies in Ge-GaAs. Phys. Rev. Lett. 40, 656–659 (1978)

42. Nelson, A.J., Gebhard, S., Kazmerski, L.L., Engelhardt, M., Höchst, H.: Formation and Schottky barrier height of Au contacts to CuInSe$_2$. J. Vac. Sci. Technol. A 9, 978–982 (1991)

43. Löher, T., Pettenkofer, C., Jaegermann, W.: Electronic properties of intimate Mo and CdS junctions in-situ formed on CuInSe$_2$ (011) cleavage planes. In Proceedings of the 24th IEEE Photovoltaic Specialists Conference, pp. 295–298 (1994)

44. Nelson, A.J., Schwerdtfeger, C.R., Wei, S.H., Zunger, A., Rioux, D., Patel, R., Höchst, H.: Theoretical and experimental studies of the ZnSe/CuInSe$_2$ heterojunction band offset. Appl. Phys. Lett. 62, 2557–2559 (1993)

45. Löher, T.: Halbleiter-Heteroübergänge: II–VI/Schichtgitterchalkogenide- und CdS/CuInSe$_2$-Grenzflächen, Dissertation, Freie Universität Berlin, (1995)

46. Niles, D.W., Contreras, M., Ramanathan, K., Noufi, R.: Determination of the valence-band offset of CdS/CIS solar cell devices by target factor analysis. In: Conference Record of the 25th IEEE Photovoltaic Specialists Conference. IEEE, New York, NY, p. 833–836 (1996)

47. Klein, A., Löher, T., Tomm, Y., Pettenkofer, C., Jaegermann, W.: Band lineup between CdS and ultra high vacuum-cleaved CuInS$_2$ single crystals. Appl. Phys. Lett. 70, 1299–1301 (1997)

48. Massé, G., Djessas, K.: X-ray photoemission studies and energy-band diagrams of (In,Se)CuInSe$_2$/SnO$_2$ heterostructures. J. Appl. Phys. 82, 825–829 (1997)

49. Wörz, M., Pschorr-Schoberer, E., Flierl, R., Preis, R., Gebhardt, W.: Photoelectron spectroscopy of chalcopyrites and Zn based II-VI semiconductor heterostructures. J. Appl. Phys. 84, 2871–2875 (1998)

50. Nadenau, V., Braunger, D., Hariskos, D., Schock, H.W.: Characterization and optimization of CuGaSe$_2$/CdS/ZnO heterojunctions. Ternary and Multinary Compounds. Proceedings of the 11th International Conference on Ternary and Multinary Compounds. Institute of Physics, Bristol, pp. 955–958 (1998)

51. Al Kuhaimi, S.A.: The electron affinity difference in CdS/CuInSe$_2$ solar cells. Sol. Energy Mater. Sol. Cells 52, 69–77 (1998)

52. Bauknecht, A., Blieske, U., Kampschulte, T., Albert, J., Sehnert, H., Lux-Steiner, M.C., Klein, A., Jaegermann, W.: Band offsets at the ZnSe/CuGaSe$_2$(001) heterointerface. Appl. Phys. Lett. 74, 1099–1101 (1999)

53. Morkel, M., Weinhardt, L., Lohmüller, B., Heske, C., Umbach, E., Riedl, W., Zweigart, S., Karg, F.: Flat conduction-band alignment at the CdS/CuInSe$_2$ thin-film solar-cell heterojunction. Appl. Phys. Lett. 79, 4482–4484 (2001)

54. Hunger, R., Pettenkofer, C., Scheer, R.: Dipole formation and band alignment at the Si(111)/CuInS$_2$ heterojunction. J. Appl. Phys. 91, 6560–6570 (2002)

55. Konovalov, I., Szargan, R.: Valence band alignment with a small spike at the CuI/CuInS$_2$ interface. Appl. Phys. Lett. 82, 2088–2090 (2003)

56. Wei, S.H., Zunger, A.: Band offsets at the CdS/CuInSe$_2$ heterojunction. Appl. Phys. Lett. 63, 2549–2551 (1993)

57. Wei, S.H., Zunger, A.: Band offsets and optical bowings of chalcopyrites and Zn-based II–VI alloys. J. Appl. Phys. 78, 3846–3856 (1995)

58. Baldereschi, A.: Mean-value point in the Brillouin zone. Phys. Rev. B 7, 5212–5215 (1973)

59. Mönch, W.: Empirical tight-binding calculation of the branch-point energy of the continuum of interface-induced gap states. J. Appl. Phys. 80, 5076–5082 (1996)

60. Penn, D.R.: Wave-Number-Dependent Dielectric Function of Semiconductors. Phys. Rev. 128, 2093–2097 (1962)

61. Tersoff, J.: Failure of the common anion rule for lattice-matched heterojunctions. J. Vac. Sci. Technol. B 4, 1066–1067 (1986)

62. Cowley, A.M., Sze, S.M.: Surface states and barrier height of metal-semiconductor systems. J. Appl. Phys. 36, 3212–3220 (1965)

63. Mönch, W.: Chemical trends of barrier heights in metal-semiconductor contacts: on the theory of the slope parameter. Appl. Surf. Sci. 92, 367–371 (1996)

64. Heske, C., Eich, D., Fink, R., Umbach, E., van Buuren, T., Bostedt, C., Terminello, L.J., Kakar, S., Grush, M.M., Callcott, T.A., Himpsel, F.J., Ederer, D.L., Perera, R.C.C., Riedl, W., Karg, F.: Observation of intermixing at the buried CdS/Cu(In,Ga)Se$_2$ thin film solar cell heterojunction. Appl. Phys. Lett. 74, 1451–1453 (1999)

65. Maierhofer, C., Kulkani, S., Alonso, M., Reich, T., Horn, K.: Valence band offset in ZnS layers on Si(111) grown by molecular beam epitaxy. J. Vac. Sci. Technol. B 9, 2238–2243 (1991)

66. Brar, B., Steibhoff, R., Seabaugh, A., Zhou, X., Jiang, S., Kirk, W.P.: Band offset measurement of the ZnS/Si(001) heterojunction. In: Melloch, M., Reed, M.A. (eds.) Proceedings of the IEEE 24th International Symposium on Compound Semiconductors, pp. 167–170. IEEE, New York (1998)

67. Mönch, W.: Barrier heights of metal contacts on H-terminated diamond: explanation by metal-induced gap states and interface dipoles. Europhys. Lett. 27, 479–484 (1994)

68. Langer, J.M., Heinrich, H.: Deep-level impurities: A possible guide to prediction of band-edge discontinuities in semiconductor heterojunctions. Phys. Rev. Lett. 55, 1414–1417 (1985)

69. Jin, M.S., Kim, W.T., Yoon, C.S.: Optical properties of undoped and co-doped CuGa$_{1-x}$Al$_x$Se$_2$ single crystals. J. Phys. Chem. Solids 54, 1509–1513 (1993)

70. Mönch, W.: In: International Workshop on Electronic, Structural, and Chemical Properties of Cu-Chalcopyrite Surfaces and Interfaces, Triberg, June 2002 (unpublished)

71. Chang, S.K., Park, H.L., Kim, H.K., Hwang, J.S., Chung, C.H., Kim, W.T.: Co^{2+} in CuGa$_{1-x}$Al$_x$Se$_2$ and Valence Band Offset in CuGa$_{1-x}$Al$_x$Se$_2$/CuGaSe$_2$ Phys. stat. sol. (b) 158, K115–K118 (1990)

72. Turcu, M., Rau, U.: Compositional trends of defect energies, band alignments, and recombination mechanisms in the Cu(In,Ga)(Se, S)$_2$ alloy system. Thin Solid Films 431–432, 158–162 (2003)

73. Sato, K., Isawa, M., Takahashi, N., Tsunoda, H.: Optical absorption spectra in CuInS$_2$ doped with Fe, Mn and Cr. Jpn. J. Appl. Phys. 27, 1359–1360 (1998)

74. Sánchez Porras, G., Wasim, S.M.: Electrical and optical properties of Fe-doped CuInSe$_2$. Phys. stat. sol. (a) 133, 509–514 (1992)

75. Rincon, C., Wasim, S.M.: In: Deb, S.K., Zunger, A. (eds.) Ternary and Multinary Compounds, p. 443. Materials Research Society, Pittsburg, PA (1997)

76. Sato, K., Tsunoda, H., Teranishi, T.: In: Deb, S.K., Zunger, A. (eds.) Ternary and Multinary Compounds, p. 459. Materials Research Society, Pittsburg, PA (1997)

77. Ruckh, M., Schmid, D., Schock, H.W.: Photoemission studies of the ZnO/CdS interface. J. Appl. Phys. 76, 5945–5948 (1994)

78. Schmid, D., Ruckh, M., Schock, H.W.: A comprehensive characterization of the interfaces in Mo/CIS/CdS/ZnO solar cell structures. Sol. Energy Mater. Sol. Cells 41/42, 281–294 (1996)

79. Hahn, B., Wörz, M., Heindel, G., Pschorr-Schoberer, E., Gerhardt, W.: Deposition and characterisation of highly conductive ZnO. Mater. Sci. Forum 287-288, 339–342 (1998)

80. Hong, S.-Ku, Hanada, T., Makino, H., Chen, Y., Ko, H.-Ju, Yao, T.: Band alignment at a ZnO/GaN (0001) heterointerface. Appl. Phys. Lett. 78, 3349–3351 (2001)
81. Sterner, J., Platzer-Björkman, C., Stolt, L.: XPS/UPS Monitoring of ALCVD ZnO Growth on Cu(In,Ga)Se$_2$ Absorbers. In: Proceedings of the 17th European Photovoltaic Solar Energy Conference, Munich, pp. 1118–1121, 2001
82. Minemoto, T., Hashimoto, Y., Satoh, T., Negami, T., Takakura, H., Hamakawa, Y.: Cu(In,Ga)Se$_2$ solar cells with controlled conduction band offset of window/Cu(In,Ga)Se$_2$ layers. J. Appl. Phys. 89, 8327–8330 (2001)
83. Platzer-Björkman, C., Lu, J., Kessler, J., Stolt, L.: Interface study of CuInSe$_2$/ZnO and Cu(In,Ga)Se$_2$/ZnO devices using ALD ZnO buffer layers. Thin Solid Films 431432, 321–325 (2003)
84. Weinhardt, L., Bär, M., Muffler, H.J., Fischer, Ch.H., Lux-Steiner, M.C., Niesen, T.P., Karg, F., Gleim, T.H., Heske, C., Umbach, E.: Impact of Cd^{2+}-treatment on the band alignment at the ILGAR-ZnO/CuIn(S, Se)$_2$ heterojunction. Thin Solid Films 431–432, 272–276 (2003)
85. Weinhardt, L., Heske, C., Umbach, E., Niesen, T.P., Visbeck, S., Karg, F.: Band alignment at the i-ZnO/CdS interface in Cu(In,Ga)(S, Se)$_2$ thin-film solar cells. Appl. Phys. Lett. 84, 3175–3177 (2004)
86. Mönch, W.: On the band structure lineup of ZnO heterostructures. Appl. Phys. Lett. 86, 122101-1-122101-3 (2005)

Defects and Self-Compensation
in Semiconductors

W. Walukiewicz

Defect formation and doping limits in semiconductors are discussed in terms of the amphoteric defect model (ADM). It is shown that the nature of defects, acceptor-like or donor-like, depends on the location of the Fermi energy relative to a common energy reference, the Fermi level-stabilization energy. The maximum free electron or hole concentration that can be achieved by doping is an intrinsic property of a given semiconductor and is fully determined by the location of the semiconductor band edges with respect to the same energy reference. The ADM provides a simple phenomenological rule that explains experimentally observed trends in free carrier saturation in a variety of semiconductor materials and their alloys. The predictions of a large enhancement of the maximum electron concentration in III–N–V alloys have been recently confirmed by experiment.

3.1 Introduction

Many of the properties of semiconductor materials are determined by native as well as foreign defects. Intentional, controlled incorporation of shallow dopants is by far the most frequent way to control conductivity of semiconductor materials. Intentional or unintentional incorporation of deep native or foreign defects is often used to produce semi-insulating semiconductor materials. It has been realized early on that many of the large variety of semiconductor materials are difficult to dope. The problem has been especially severe in wide-bandgap semiconductors where in many instances n-type or p-type doping cannot be achieved at all, significantly limiting the range of applications of these materials [1–3].

The past several years have witnessed spectacular progress in the development of a new generation of short-wavelength optoelectronic devices based on group III nitrides [4–6] and wide-gap II–VI semiconductors [7,8]. In both cases, this progress was made possible through the discovery of more efficient ways to activate acceptor impurities in these material systems. Despite this progress, the high resistance of p-type layers is still a major hurdle in the development of the devices requiring high current injection levels.

There have been numerous attempts to understand the maximum doping limits in semiconductors. Most of these were aimed at explaining limitations imposed on a specific dopant in a specific semiconductor. Thus, it has been argued that in the case of amphoteric impurities in III–V compounds, doping is limited by the impurities occupying both acceptor and donor sites, compensating each other. Redistribution of impurities can also lead to limitations of the maximum doping level in the materials with impurity diffusion strongly depending on the Fermi energy [3]. Formation of new stable solid phases involving dopant atoms can be a severe limitation in achieving high doping levels [9].

Passivation of donor and acceptor impurities by highly mobile impurities is another major mechanism limiting the electrical activity of dopants. Hydrogen, lithium, and copper are known to passivate intentionally introduced dopants in semiconductors. Hydrogen has been an especially extensively studied impurity as it is a commonly used element in most semiconductor processing techniques and in all the growth techniques involving metal organic precursors [10]. In some cases hydrogen can be removed during postgrowth annealing. Magnesium-doped p-type GaN is frequently obtained by thermal annealing of metal organic chemical vapor deposition (MOCVD) grown, hydrogen passivated films [11, 12]. However, in other instances, as in the case of N-doped ZnSe, hydrogen is too tightly bound to the N acceptors and cannot be removed by thermal annealing [13].

Over the last few years considerable effort has been directed towards overcoming the doping limits. For example, it has been proposed that one can enhance incorporation of electrically active centers by co-doping with donors and acceptors. It has been argued based on theoretical calculations that because of the reductions of the lattice relaxation and Madelung energies, formation energies of proper donor acceptor complexes can be lower than the formation energy of isolated dopant species [13]. Some preliminary experimental results indicate that indeed the co-doping method has produced p-type ZnO that cannot be achieved by any other method [14]. Further studies are needed to fully understand the issues of poor reproducibility of the results obtained by the co-doping method.

In this chapter, the formation of defects and saturation of the free carrier concentration in semiconductors will be discussed in terms of the ADM. In recent years, the model has been successfully applied to numerous doping-related phenomena in semiconductors. It has been used to explain doping-induced suppression of dislocation formation [15] as well as impurity segregation [16, 17] and interdiffusion [18] in semiconductor superlattices. We will show that ADM provides a simple phenomenological rule capable of predicting trends in the nature of the defects and the doping behavior of a large variety of semiconductor systems.

3.2 Fermi-Level Stabilization Energy

All point defects and dopants can be divided into two classes: delocalized, shallow dopants and highly localized defects and dopants. Shallow hydrogenic donors and acceptors belong to the first class. Their wave functions are delocalized and formed mostly out of the states close to the conduction band minimum or the valence band maximum. As a result, the energy levels of these dopants are intimately associated with the respective band edges, conduction band for donors and the valence band for acceptors. In general, the energy levels follow the respective band edges when the locations of the edges change due to external perturbation such as hydrostatic pressure or changing alloy composition.

In contrast, wave functions of highly localized defects or dopants cannot be associated with any specific band structure extremum. They are rather, formed from all the extended states in the Brillouin zone with the largest contribution coming from the regions of large density of states in the conduction and valence bands. Consequently, the energy levels of such defects or dopants are insensitive to the location of the low density of states at the conduction and valence band edges. For example, it has been shown that transition metal impurities with their highly localized d shells belong to this class of dopants [19, 20]. The insensitivity of the transition metal energy levels to the position of local band extrema has led to the concept of using these levels as energy references to determine the band offsets in III–V and II–VI compounds [20] and the band edge deformation potentials in GaAs and InP [21].

Compelling evidence for the localized nature of native defects has been provided by studies of semiconductor materials heavily damaged with high gamma rays or electrons [22–28]. It has been found that for sufficiently high damage density, i.e., when the properties of the material are fully controlled by native defects, Fermi energy stabilizes at a certain energy and becomes insensitive to further damage. As is shown in Fig. 3.1, in GaAs a high-energy electron damage leads to stabilization of the Fermi energy at about 0.6 eV above the valence band edge [23]. The location of this Fermi level-stabilization energy, E_{FS}, does not depend on the type or the doping level of the original material and therefore is considered to be an intrinsic property of a given material. As is shown in Fig. 3.2, the Fermi level-stabilization energies for different III–V semiconductors line up across semiconductor interfaces and are located approximately at a constant energy of about 4.9 eV below the vacuum level [29]. This is a clear indication that the native defect states determining the electrical characteristics of heavily damaged materials are of highly localized nature. As can be seen in Fig. 3.2, the location of the stabilized Fermi energy in heavily damaged III–V semiconductors is in good agreement with the Fermi level-pinning position observed at metal/semiconductor interfaces [31]. This finding strongly supports the assertion that the same defects are responsible for the stabilization of the Fermi energy in both cases.

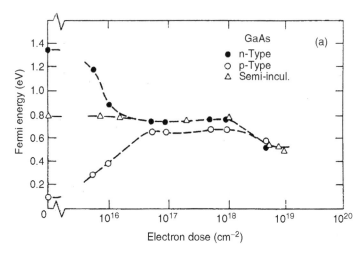

Fig. 3.1. Dependence of Fermi energy on high-energy electron irradiation dose based on the results presented in [23]

3.3 Amphoteric Native Defects

The mechanism explaining the defect-induced stabilization of the Fermi energy is based on the concept of amphoteric native defects. The stabilization of the Fermi energy can be understood if we assume that the type of defects formed during high-energy particle irradiation or metal deposition on the semiconductor surface depends on the location of the Fermi energy with respect to E_{FS}. For Fermi energy, $E_F > E_{FS}$ ($E_F < E_{FS}$), acceptor-like (donor-like) defects are predominantly formed resulting in a shift of the Fermi energy towards E_{FS}. Consequently, the condition $E_F = E_{FS}$ is defined as the situation where the donor-like and acceptor-like defects are incorporated at such rates that they perfectly compensate each other leaving the Fermi energy unchanged.

Such an amphoteric behavior of simple native defects is supported by theoretical calculations that show that depending on the location of the Fermi energy vacancy like defects can acquire either negative or positive charge acting as acceptors or donors, respectively. In the case of GaAs it was shown that both gallium and arsenic vacancies could undergo amphoteric transformations [32]. For example, as shown in Fig. 3.3, V_{Ga} is a triple acceptor for $E_F > E_v + 0.6\,\text{eV}$. However, for lower Fermi energies this configuration is unstable and the vacancy undergoes a relaxation, in which one of the first neighboring As atoms moves towards the vacant Ga site. The transformation is schematically represented by the reaction

$$V_{Ga} \rightarrow (V_{As} + As_{Ga}). \tag{3.1}$$

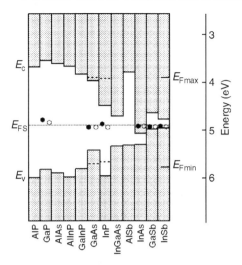

Fig. 3.2. Band offsets, the Fermi level-stabilization energy (E_{FS}), and the extremum Fermi level positions (E_{Fmax} and E_{Fmin}) in III–V compounds. The energy is measured relative to the vacuum level. The *filled circles* represent stabilized Fermi energies in heavily damaged materials, exposed to high-energy radiation. The *open circles* correspond to the location of the Fermi energy on pinned semiconductor surfaces and at metal/semiconductor interfaces [29]. The *dashed lines* present the Fermi level positions obtained from the maximal experimentally observed electron and hole concentrations [30]

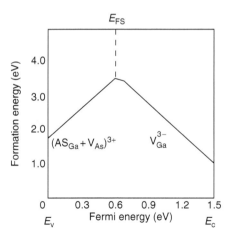

Fig. 3.3. Formation energy of a gallium vacancy and the related donor defect as function of the Fermi energy in the GaAs band gap [32, 33]

In arsenic-rich GaAs the calculated formation energy of V_{Ga} is below 1 eV for E_F at the conduction band edge [33].

A similar amphoteric behavior is also predicted for V_{As} where the transformation is given by the reaction [32]

$$V_{As} \rightarrow (Ga_{As} + V_{Ga}). \tag{3.2}$$

In this case, it is the V_{As} donor that is stable in GaAs with E_F larger than about $E_v + 0.8\,eV$ and transforms to an acceptor-like ($V_{Ga} + Ga_{As}$) configuration for $E_F < E_v + 0.8\,eV$ [27]. It is worth noting that these theoretical values of E_{FS} are very close to experimentally determined ones, ranging from $E_v + 0.5\,eV$ to $E_v + 0.7\,eV$ [23].

Most recent theoretical calculations have shown that the amphoteric behavior of native defects is a feature common to many different compound semiconductor systems, including II–VI and III–V semiconductors and the group III nitrides [34]. The calculations have confirmed that reaction (3.1) is responsible for the amphoteric behavior of V_{Ga}. However, it has been found that in the case of V_{As}, a transformation from a donor-like V_{As} to an acceptor-like configuration occurs through a dimerization of the threefold coordinated Ga atoms surrounding the As vacancy rather than reaction 3.2. Although a different type of structural relaxation is predicted in this case, it does not change the overall conclusion that both cation and anion site vacancies are amphoteric defects and, when introduced in large concentrations, will lead to a stabilization of the Fermi energy.

Since E_{FS} is associated with highly localized defects, its location is not correlated with the positions of the conduction or valence band edges. Thus, as can be seen in Fig. 3.2, E_{FS} can be located anywhere in the gap or even in the conduction band. In the case of GaAs, E_{FS} is located close to the midgap energy. Therefore, high-energy radiation damage always leads to a high-resistivity GaAs [23]. On the other hand, in the unusual case of InAs, E_{FS} is located deep in the conduction band. Consequently, any high-energy radiation damage leads to high n-type conductivity in this material [26]. It has been shown that the location of E_{FS} relative to the band edges is the single, most important factor affecting defect-related phenomena in semiconductors.

3.4 Maximum Doping Limits in GaAs

It has been realized very early that it is rather easy to dope GaAs with acceptors. Very high hole concentrations, in excess of $10^{20}\,cm^{-3}$, can be readily obtained by doping with group II atoms [35]. Even higher concentrations close to $10^{21}\,cm^{-3}$ were obtained by doping with carbon [36]. On the other hand, n-type doping is much more difficult to achieve. The doping becomes less efficient for donor concentrations larger than about $3 \times 10^{18}\,cm^{-3}$ and the maximum electron concentration saturates at a level slightly above $10^{19}\,cm^{-3}$

[37–40]. The maximum concentration does not depend on the dopant species or the method by which the dopants are introduced into the crystal. Therefore, this limitation appears to be an intrinsic property of the material, rather than a feature attributable to the chemical or electronic characteristics of the dopants.

Over the years numerous attempts were made to understand the nature of this limitation. For example, it has been proposed that at high concentrations Se donors form electrically inactive complexes [37]. In the case of group IV dopants, an obvious explanation was based on the amphoteric nature of these impurities. It was argued that at high doping levels the dopants begin to occupy both sites forming donors and acceptors that compensate each other [40]. It would be rather surprising if these dopant-specific explanations could account for the universal nature of the electron concentration limits.

These results point at the intrinsic nature of the mechanism limiting the free electron concentration in GaAs. Calculations of the electron concentration as function of the doping levels were performed assuming that triple negatively charged V_{Ga} are responsible for the compensation [18]. The results are shown in Fig. 3.4. A good fit to experimental data was obtained assuming that the formation energy of $V_{Ga}E_f = 2.4\,eV$ for the Fermi energy located at the intrinsic level. As is seen in Fig. 3.4, the results of the calculations reflect quite well the overall dependence of the electron concentration on the doping level, N_d. At low N_d the Fermi energy is located well below the conduction band, E_f is large and the concentration of V_{Ga} small. Under these conditions all donors are electrically active and $n = N_d$. The doping induced upward shift of the Fermi energy towards the conduction band results in a lower E_f and a higher $[V_{Ga}]$. Gallium vacancies compensate the donors and the electron concentration becomes a sublinear function of N_d. In fact, it can be shown that in a limited concentration range, n is proportional to $(N_d)^{1/3}$. The 1/3 power dependence reflects the fact that V_{Ga} is a triply charged acceptor. Such dependence is expected when electrons can still be described by nondegenerate statistics. At even higher doping levels, the Fermi energy enters the conduction band and becomes strongly dependent on electron concentration [41]. This leads to a rapid reduction of E_f, an increase of V_{Ga}, and as a consequence saturation of n.

It is important to note that the value of $E_f = 2.4\,eV$ appears to be consistent with other determinations of the formation energy of V_{Ga} in intrinsic GaAs. Detailed studies of Ga self-diffusion in undoped GaAs provided the value of the diffusion activation energy, which is a sum of the formation and migration energies of V_{Ga}, $E_{f+m} = E_f + E_m = 3.7\,eV$ [42]. The entropy of $S = 3.5k$ has also been determined in this study. In addition, extensive investigations of V_{Ga} facilitated diffusion of As_{Ga} defects in non-stoichiometric, low-temperature-grown GaAs have provided the values of V_{Ga} migration energies ranging from 1.4 to 1.7 eV [43]. This leads to E_f ranging from 2.0 to 2.3 eV, which is somewhat lower than the value of $E_f = 2.4\,eV$ needed to explain the free electron concentration limits. The difference can easily be

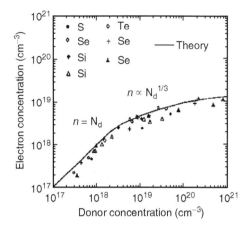

Fig. 3.4. Electron concentration as a function of donor doping in GaAs. The data points represent experimental results for several different donor species (*filled circles*—S [38], *open circles*—Se [38], *filled diamonds*—Si [37], *open triangles*—Si [40], *open diamonds*—Te [38], *plus sign*—Se [37], and *filled triangles*—Se [39])

accounted for by the entropy contribution that has been neglected in the present considerations. At 900 K, the entropy of 3.5 k leads to an effective formation energy difference of about 0.27 eV.

The success in explaining the doping limitations in n-type GaAs raises the question whether a similar mechanism is responsible for doping limits in p-type GaAs. As is shown in Fig. 3.3, V_{Ga} is an unstable defect for $E_F < E_{FS}$. It relaxes to the $V_{As}+As_{Ga}$ donor-like configuration with the formation energy $E_f = E_{f0} + 3(E_F - E_{FS})$. With $E_{f0} = E_f(E_{FS}) = 3.1$ eV, one finds that at a temperature of $T = 900$ K for E_F located at the valence band edge E_v, the formation energy, $E_f = 1.8$ eV. This large formation energy gives a very low value of less than 10^{13} cm^{-3} for the concentration of the defect donors. Since for $E_F = E_v$ the concentration of free holes is equal to about 4×10^{19} cm^{-3}, it is evident that $(V_{As} + As_{Ga})$ donors are not expected to play any role in the compensation of intentionally introduced acceptors. This is consistent with experiments that indicate that rather high hole concentrations can be relatively easily achieved in p-type GaAs.

However, it has also been shown that in GaAs doped with column II acceptors the hole concentration saturates at the doping levels slightly above 10^{20} cm^{-3} [44]. This saturation has been attributed to the fact that column II atoms can act either as acceptors when they substitute Ga atom sites, or as donors when they occupy interstitial sites. The concentration ratio of substitutional to interstitial atoms depends on the location of the Fermi energy. At low concentrations all dopant atoms substitute Ga sites acting as acceptors. With increasing doping level the Fermi energy shifts down towards the valence band and more and more dopants occupy interstitial sites acting as donors. As has been shown earlier [16, 41], this mechanism leads to a saturation of the position of the Fermi energy level and thus also of the concentration of free holes in the valence band. In the case of GaAs, with the maximum hole concentration of $\sim 10^{20}$ cm^{-3}, the Fermi energy saturates at about $E_v - 0.2$ eV or at $E_{FS} - 0.67$ eV when measured with respect to E_{FS} as a common energy reference.

The extremum Fermi level positions, E_{Fmax} and E_{Fmin}, corresponding to the saturation levels of hole and electron concentrations in GaAs and other III–V semiconductors, are also shown in Fig. 3.2. Like the Fermi-level stabilization energy, the extremum Fermi-level positions also align on an absolute energy scale. They can be used to explain and predict doping limits in the whole class of semiconductors.

3.5 Group III Nitrides

Recent years have witnessed an unprecedented growth of interest in the group III nitrides as a new distinct class of III–V compounds with strongly ionic bonds, smaller lattice constants and large bandgaps. These materials form the foundation of a new technology for short-wavelength optoelectronics [4] and high-power, high-speed electronic devices [45]. Although the group III nitrides have been studied for many years, it has been discovered only recently that the energy gap of InN is only 0.7 eV, which is much smaller than the previously accepted value of 1.9 eV [46, 47]. This lower energy gap of InN has led to a re-evaluation of the bandgap bowing parameters and the band offsets in group III nitride alloys [48]. These alloys show an unprecedented large range of the direct band-gap energies; from near-infrared in InN to deep ultraviolet in AlN [48].

The large conduction band offsets lead to interesting trends in the doping behavior of the nitride alloys. As is shown in Fig. 3.5, in InN E_{FS} is located deeply in the conduction band at $E_c + 0.8$ eV. This explains the extreme propensity of this material to n-type doping. To date no p-type conducting InN has been realized yet. At the same time, InN with electron concentrations as high as 10^{21} cm^{-3} are readily available [49]. GaN with the E_{FS} located in the upper half of the bandgap is a good n-type conductor with electron concentrations in excess of 10^{20} cm^{-3} whereas hole concentration is limited to about 10^{18} cm^{-3} in this material.

3.6 Group III–N–V Alloys

An excellent example for the predictive power of the ADM has been the recently discovered high activation efficiency of shallow donors in GaInNAs alloys. It has been shown more than 10 years ago that alloying of group III–V

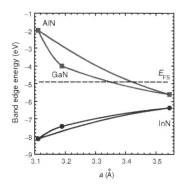

Fig. 3.5. Band offsets for group III nitrides. The *dashed lines* represent the Fermi-level stabilization energy. The energy scale is referenced to the vacuum level

compounds with group III nitrides leads to dramatic change of the electronic properties of the resulting group III–N–V alloys [50]. For example, GaNAs with only 1% of N has its bandgap reduced by 0.18 eV [51]. We have shown recently that the reduction of the bandgap results from an interaction between highly localized nitrogen states and the extended states of the host semiconductor matrix [52, 53]. The interaction splits the conduction band into two subbands with highly non-parabolic dispersion relations. It has been shown that the dispersion relation for the lower E_- and upper E_+ conduction subbands are given by

$$E_{\pm}(k) = \{(E_N + E_M(k)) \pm [(E_N - E_M(k))^2 + 4(V_{NM})^2]^{1/2}\}/2, \qquad (3.3)$$

where $E_M(k)$ is the conduction band energy of the host semiconductor matrix, E_N is the energy of the localized nitrogen levels and V_{NM} is the coupling parameter. For a random distribution of N atoms, $V_{NM} = C_{NM}x^{1/2}$; where x is the molar fraction of N atoms in the alloy and C_{NM} is a constant dependent on the host semiconductor material only. The downward shift of the lower conduction subband given by Eq. 3.2 accounts for the large reduction of the fundamental bandgap [53]. Also, the interaction leads to a large enhancement of the electron effective mass and thus also to an increased density of states of the lower conduction subband [54].

Figure 3.6 shows the location of the conduction band edge as function of the N content in GaN$_x$As$_{1-x}$. The Fermi-level stabilization energy is also shown in this figure. It is seen that both the downward shift of the conduction band edge and the increase in the density of states of the lower subband should result in a higher maximum electron concentration in GaNAs alloys [53]. Studies of Se-doped GaInNAs alloys have fully confirmed these predictions [55]. As is shown in Fig. 3.7, GaInNAs alloys with a relatively small N content exhibit a large enhancement of the maximum electron concentration. A more than one order of magnitude improvement of donor activation efficiency has been found in GaInNAs with only 3.3% N.

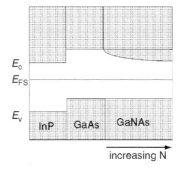

Fig. 3.6. Schematic representation of the conduction band and the valence band alignment of GaNAs compared with GaAs and InP

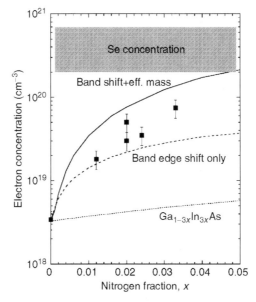

Fig. 3.7. A comparison of the measured and calculated maximum free electron concentrations as functions of the N content in $Ga_{1-3x}In_{3x}N_xAs_{1-x}$. Two different cases for calculated n_{max} are shown: one includes effects of the downward shift of the conduction band edge only (*dashed curve*) and the other includes both the band shift and the enhancement of the density of states effective mass (*solid curve*). The *dotted line* shows the increase in n_{max} expected in $Ga_{1-3x}In_{3x}As$ alloys. The *shaded area* indicates the range of Se concentration in the studied samples

Our most recent studies indicate that the band anticrossing model well describes the electronic structure of a broad class of highly mismatched semiconductor alloys. We have shown that in addition to III–N–V alloys, large downward shifts of the conduction band are also found in group II–VI alloys such as ZnS_xTe_{1-x} or $ZnSe_yTe_{1-y}$ in which the element Te is partially replaced by much more electronegative S or Se [56]. It is therefore expected that one could significantly improve the donor activation efficiency by alloying ZnTe with ZnS or ZnSe. For example, a more than one order of magnitude higher maximum electron concentration is expected in ZnSTe with only few percent of S. Experimental evidence is available for ZnSeTe: In ZnSe the p-dopability has been greatly improved by adding 3% of ZnTe [57].

3.7 Group II–VI Semiconductors

Wide-gap group II–VI semiconductors are the group of materials that exhibit the most severe limitations on doping. Indeed, it is this family of materials for which the problem of doping limits has been recognized first [2]. Early studies have shown that all wide-gap II–VI compounds show a propensity for either n-type or p-type conductivity. As-grown ZnO, ZnS, HgSe, CdSe and CdS show n-type conductivity; p-type doping is very difficult if not impossible to achieve in these compounds. On the other hand, ZnTe typically exhibits p-type conductivity only. It was recognized at that time that the doping limits could originate from compensating native defects that are formed when the Fermi energy shifts towards the band edges [2]. It was not clear, however, how within this picture one could explain differences between apparently similar materials exhibiting completely different doping behavior.

The work on the utilization of II–VI compounds for short-wavelength light-emitting devices have brought the issue of the doping limitations to the fore-front and led to intensive efforts aimed at understanding the mechanisms responsible for the limited dopability of these materials [58, 59]. Because of its importance for the blue-green light emitters, ZnSe has been considered a prototypical material to study the doping limitations. It can be relatively easily doped n-type, but p-type doping is very difficult to accomplish and only recently doping with reactive nitrogen was successful in achieving p-type conductivity. However, even in this case the free-hole concentration is limited to $10^{18}\,\mathrm{cm}^{-3}$ [58].

One explanation for this effect is based on the argument that it is energetically favorable for the dopant species to form new compounds with the host crystal atoms rather than substitute lattice sites and act as donors or acceptors [16]. In the case of N-doped ZnSe, the calculations suggested that Zn_3N_2 should be easily formed preventing N from acting as a substitutional acceptor [9]. Also, these first-principle calculations seemed to indicate that the formation energies of native defects are too large and the concentrations are too small to explain low electrical activity of N atoms in ZnSe with compensation by native defects [9]. Later, improved calculations have shown that incorporation of lattice relaxation lowers the formation energy of native defects so that they are likely to play a role in the compensation of N acceptors in ZnSe [60].

Native defects were frequently invoked as the centers compensating electrical activity of intentionally introduced dopants. It was very difficult, however, to identify the defects responsible for the compensation or to account for the trends in the doping behavior observed in different II–VI compounds and their alloys. There is evidence that in the specific case of ZnSe:N, V_{Se} or V_{Se}–N defect complexes are responsible for the compensation of p-type conduction [61]. This finding, however, does not provide any guidance on how to identify the compensating defects in other II–VI compounds.

It has been shown that the trends in the doping behavior of different group II–VI compounds can be understood within the ADM without any need to know the specific identity of the compensating defects [62,63]. The conduction and valence bands for various II–VI semiconductors are shown in Fig. 3.8 [62]. Concluding from the observations in III–V materials the Fermi level-stabilization energy is expected to be located again at about 4.9 eV below the vacuum level. As in the case of III–V compounds, it is assumed that there is a band of allowed Fermi energies, $\Delta E_F = E_{Fmax} - E_{Fmin}$, around E_{FS} determining the maximum electron and hole concentrations that can be achieved in a given material.

In the case of ZnSe, the highest electron concentration of about 2×10^{19} cm^{-3} [64] defines $E_{Fmax} = E_{FS} + 1.3$ eV as the upper limit of allowed Fermi energies. The lower limit at $E_{FS} - 1.3$ eV corresponds to a maximum free-hole concentration of 10^{18} cm^{-3} [58]. Transferring the same limits to other compounds, we find that in ZnTe E_{Fmin} is located deep in the valence band, confirming the experimental observation that it is very easy to dope this material with acceptors. Indeed, free-hole concentrations as high as 10^{20} cm^{-3} were reported in ZnTe [65]. On the other hand, since E_{Fmax} is located below the conduction band edge, it is expected that n-type conductivity will be much more difficult to achieve. In fact, it was only recently that n-type conduction with a low electron concentration of 4×10^{17} cm^{-3} was reported in ZnTe [66].

As can be seen in Fig. 3.8 for CdSe and CdS, the upper Fermi energy limit is located in the conduction band in agreement with the observation that both materials are very good n-type conductors. As expected p-type conductivity is much more difficult to realize in these materials. A maximum hole concentration of only 10^{17} cm^{-3} was reported in CdSe [67]. It is not surprising that in CdS with its very low position of the valence band no p-type doping was ever achieved.

ZnO represents a case of a material with the band edges shifted to very low energies. The conduction band edge is located very close to E_{FS} at $E_{FS}+0.2$ eV and the valence band edge lies at the very low energy of $E_{FS} - 3.1$ eV. Such an alignment strongly favors n-type conductivity. Existing experimental data indicate that undoped ZnO can exhibit free electron concentrations as large as 1.1×10^{21} cm^{-3} [68]. However, the extremely low position of the valence band edge indicates that it will be very difficult, if not impossible, to achieve any p-type doping of this material.

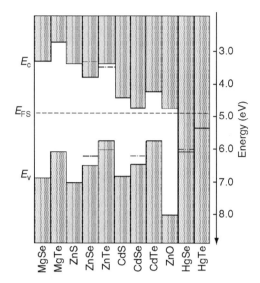

Fig. 3.8. Band offsets and the Fermi-level stabilization energy, E_{FS}, in II–VI compounds. The *dashed lines* represent positions of the Fermi energy corresponding to the highest hole and electron concentrations reported for the given material, E_{Fmax}, E_{Fmin}

3.8 Group I–III–VI$_2$ Chalcopyrites

As has been shown above, the ADM has been successfully applied to explain the defect behavior in group III–V and group II–VI compounds and their alloys. The question arises whether the same approach could provide any guidance on the defect behavior in more complex compounds. Group I–III–VI$_2$ ternary chalcopyrites represent an important class of semiconductor materials that have been extensively studied for more than three decades. Special attention has been devoted to Cu-chalcopyrites such as CuInSe$_2$ and CuGaSe$_2$ as they are key component materials in the design of efficient, radiation-hard solar cells [69, 70].

It has been argued that the energies of deep native defects remain constant on the absolute energy scale and therefore can be used to determine the band offsets between different compounds [71]. This indicates that like in III–V or II–VI compounds, one could use the ADM to explain the doping behavior in these more complex compounds. Indeed, several years ago a phenomenological approach similar to the ADM has been applied to address the issues of the free carrier concentration limits in I–III–VI$_2$ semiconductors [72]. A reasonably good correlation has been found between the location of the conduction or the valence band edges and the propensity for a specific type of conductivity. For example, the group I–III–Te$_2$ materials that have the highest location of the valence band edges show very high hole concentrations. It should be noted, however, that these materials are much more complex than simple binary compounds. They are much more difficult to dope with impurities. In most instances, deviations from stoichiometry are used to control the type of doping [73]. It is clear that much more systematic work on the properties of defects will be required to understand the applicability of the amphoteric defect concept to the group I–III–VI$_2$ ternaries better.

3.9 Conclusions

It has been shown that native defects in a semiconductor crystal lattice exhibit amphoteric behavior. Depending on the location of the Fermi energy they can act either as acceptors or donors. The demarcation Fermi energy separating donor-like from acceptor-like behavior, plays an important role of the energy at which the Fermi level is stabilized in the presence of large concentrations of native defects. It also serves as a convenient energy reference to evaluate the Fermi energy-dependent part of the defect formation energy. Based on these observations, a model has been developed that addresses the issue of the relationship between the native defects and intentionally introduced dopants. It is shown that the maximum free electron or hole concentration that can be achieved by doping is an intrinsic property of a given semiconductor and is fully determined by the location of the semiconductor band edges with respect to the Fermi level-stabilization energy. The ADM provides a simple phenomenological rule that explains experimentally observed trends in free carrier saturation in semiconductors. It correctly describes the maximum attainable concentrations of free electrons and free holes in a variety of semiconductor materials systems. It has also been used successfully in addressing other issues including impurity segregation and interdiffusion in semiconductor heterostructures and doping-induced suppression of dislocation formation.

Use of complex, layered structures of different semiconductor materials plays an increasingly important role in the design of modern optoelectronic devices. Such structures allow not only to tune the emitted light energy but also to control the confinement and separation of free electron and hole systems. This is achieved by the proper tuning of the conduction and the valence band offsets between different component layers of the devices. The problems of the maximum doping and impurity redistribution within such device structures have always been treated as entirely separate issues. The ADM unifies those two apparently unrelated aspects of optoelectronic devices by providing a simple rule relating the maximum doping levels and dopant diffusion and redistribution to the same conduction and valence band offsets that control the distribution of free electrons and holes in optoelectronic devices.

Acknowledgments

This work was supported by the Director, Office of Science, Office of Basic Energy Sciences, Division of Materials Sciences and Engineering, of the U.S. Department of Energy under Contract No. DE-AC03-76SF00098.

References

1. Longini, R.L., Greene, R.F.: Ionization interaction between impurities in semi-conductors and insulators. Phys. Rev. 102, 992–999 (1956)
2. Kroger, F.A., Vink, H.J.: Relations between the Concentrations of Imperfections in crystalline Solids. Solid State Phys. III, 307–435 (1956)
3. Schubert, E.F.: Doping in III-V Semiconductors. Cambridge University Press, Cambridge (1993)
4. Nakamura, S.: InGaN/AlGaN blue-light-emitting diodes. J. Vac. Sci. Technol. A 13, 705–710 (1995)
5. Akasaki, I., Sota, S., Sakai, H., Tanaka, T., Koike, M., Amano, H.: Shortest wavelength semiconductor laser diode. Electron. Lett. 32, 1105–1106 (1996)
6. Perlin, P., Gorczyca, I., Christansen, N.E., Grzegory, I., Teisseyre, H., Suski, T.: Pressure studies of gallium nitride: Crystal growth and fundamental electronic properties. Phys. Rev. B 45, 13307–13313 (1992)
7. Park, R.M., Troffer, M.B., Rouleau, C.M., DePuydt, J.M., Haase, M.A.: p-type ZnSe by nitrogen atom beam doping during molecular beam epitaxial growth. Appl. Phys. Lett. 57–59, 2127 (1990)
8. Ishibashi, A.: II–VI blue–green light emitters. Proceedings of the 7th International Conference on II-VI Compounds and Devices, Edinburgh, Scotland, UK, August, pp. 555–565. North Holland, Amsterdam (1995)
9. Laks, D.B., Van de Walle, C.G., Neumark, G.F., Blochl, P.E., Pantelides, S.T.: Native defects and self-compensation in ZnSe. Phys. Rev. B 45, 10965–10978 (1992)
10. Haller, E.E.: Hydrogen in crystalline semiconductors. Semicond. Sci. Technol. 6, 73–84 (1991)
11. Akasaki, I., Amano, H.: Crystal growth and conductivity control of group III nitride semiconductors and their application to short wavelength light emitters. Jpn. J. Appl. Phys. 36, 5393–5408 (1997)
12. Nakamura, S., Senoh, M., Mukai, T.: Highly P-typed Mg-doped GaN films grown with GaN buffer layers. Jpn. J. Appl. Phys. 30, L1708–1710 (1991)
13. Yamamoto, T., Yoshida, H.K.: Solution using a codoping method to unipolarity for the fabrication of p-type ZnO. Jpn. J. Appl. Phys. 38, L166–L168 (1999)
14. Joseph, M., Tabata, H., Kawai, T.: P-type electrical conduction in ZnO thin films by Ga and N codoping. Jpn. J. Appl. Phys. 38, L1205–L1207 (1999)
15. Walukiewicz, W.: Doping-induced suppression of dislocation formation in semi-conductors. Phys. Rev. B 39, 8776–8779 (1989)
16. Walukiewicz, W.: Application of the amphoteric native defect model to diffusion and activation of shallow impurities in III-V semiconductors. Mater. Res. Soc. Symp. Proc. 300, 421–432 (1993)
17. Walukiewicz, W.: Activation and diffussion of shallow impurities in compound semiconductors. Inst. Phys. Conf. Ser. 141, 259–264 (1994)
18. Walukiewicz, W.: Amphoteric native defects in semiconductors. Appl. Phys. Lett. 54, 2094–2097 (1989)
19. Zunger, A.: Theory of 3d transition atom impurities in semiconductors. Ann. Rev. Mater. Sci. 15, 411–453 (1985)
20. Langer, J.M., Heinrich, H.: Deep-level impurities: A possible guide to prediction of band-edge discontinuities in semiconductor heterojunctions. Phys. Rev. Lett. 55, 1414–1417 (1985)

21. Nolte, D.D., Walukiewicz, W., Haller, E.E.: Band-edge hydrostatic deformation potentials in III–V semiconductors. Phys. Rev. Lett. 59, 501–504 (1987)
22. Brudnyi, V.N., Novikov, V.A.: Limiting electrical parameters of electron-irradiated gap. Fiz. Tekh. Poluprovodn. 19, 747–748 (1985); Sov. Phys. Semicond. 19, 460–461 (1985)
23. Brudnyi, V.N., Krivov, M.A., Potapov, A.I., Shakhovostov, V.I.: Electrical properties and defect annealing in Gallium-Arsenide irradiated with large electron doses. Fiz. Tekh. Poluprovodn. 16, 39–42 (1982); Sov. Phys. Semicond. 16, 21–23 (1982)
24. Cleland, J.W., Crawford, J.H.: Fast-neutron bombardment of GaSb. Phys. Rev. 100, 1614–1618 (1955)
25. Brudnyi, V.N., Novikov, V.A.: 'Limiting' electrical properties of irradiated InP. Fiz. Tekh. Poluprovodn. 16, 1880–1881 (1982); Sov. Phys. Semicond. 16, 1211–1212 (1982)
26. Kekelidze, N.P., Kekelidze, G.P.: Electrical and optical properties of InAs and InP compounds and their solid solutions of InP_xAs_{1-x} irradiated with fast neutrons and gamma rays. In: Radiation Damage and Defects in Semiconductors, Institute of Physics Conference Series, vol. 16, pp. 387–393. Institute of Physics, London (1972)
27. Mashovets, T.V., Khansevarov, R.Yu.: Low-temperature gamma irradiation and annealing of Indium Antimonide. Fiz. Tverd. Tela 8, 1690–1691 (1966); Sov. Phys. Solid State 8, 1350–1351 (1966)
28. Walukiewicz, W.: Mechanism of Fermi-level stabilization in semiconductors. Phys. Rev. B 37, 4760–4763 (1988)
29. Walukiewicz, W.: Fermi level dependent native defect formation: consequences for metal-semiconductor and semiconductor–semiconductor interfaces. J. Vac. Sci. Technol. B 6, 1257–1262 (1988)
30. Tokumitsu, E.: Correlation between Fermi level stabilization positions and maximum free carrier concentrations in III–V compound semiconductors. Jpn. J. Appl. Phys. 29, L698–L700 (1990)
31. Walukiewicz, W.: Mechanism of Schottky barrier formation: the role of amphoteric native defects. J. Vac. Sci. Technol. B 5, 1062–1067 (1987)
32. Baraff, G.A., Schluter, M.S.: Electronic structure, total energies, and abundances of the elementary point defects in GaAs. Phys. Rev. Lett. 55, 1327–1330 (1985)
33. Zhang, S.B., Northrup, J.E.: Chemical potential dependence of defect formation energies in GaAs: Application to Ga self-diffusion. Phys. Rev. Lett. 67, 2339–2342 (1991)
34. Chadi, D.J.: Inverted charge states of anion and cation-site vacancies in zinc blende semiconductors: theory. Materials Science Forum, vols. 258-263, p. 1321 (1997)
35. Schubert, E.F., Kuo, J.M., Kopf, R.F., Luftman, H.S., Hopkins, L.C., Sauer, N.J.: Beryllium delta doping of GaAs grown by molecular beam epitaxy. J. Appl. Phys. 67, 1969–1979 (1990)
36. Yamada, T., Tokumitsu, E., Saito, K., Akatsuka, T., Miyauchi, M., Konagai, M., Takahashi, K.: Heavily carbon doped para-type GaAs and GaAlAs grown by metalorganic molecular-beam epitaxy. J. Cryst. Growth 95, 145–149 (1989)
37. Vieland, L.J., Kudman, I.: Behavior of selenium in gallium arsenide. J. Phys. Chem. Solids 24, 437–439 (1963)

38. Milvidskii, M.G., Osvenskii, V.B., Fistul, V.I., Omelyanovskii, E.M., Grishina, S.P.: Influence of heat treatment on electrical properties of heavily doped n-type Gallium Arsenide. Sov. Phys. Semicond. 1, 813–817 (1968)
39. Emelyanenko, O.V., Lagunova, T.S., Nasledov, D.B.: Scattering of current carriers in Gallium Arsenide in the presence of strong degeneracy. Sov. Phys. Solid State 2, 176–180 (1960)
40. Whelan, J.M., Struthers, J.D., Ditzenberger, J.A.: Effects of silicon in gallium arsenide. In Proceedings of the International Conference on Semiconductor Physics, Prague, 1960, pp. 943–945. Academic Press, New York (1961)
41. Walukiewicz, W.: Defect formation and diffusion in heavily doped semiconductors. Phys. Rev. B 50, 5221–5225 (1994)
42. Bracht, H., Haller, E.E., Eberl, K., Cardona, M.: Self- and interdiffusion in $Al_x/Ga_{1-x}/As/GaAs$ isotope heterostructures. Appl. Phys. Lett. 74, 49–51 (1999)
43. Bliss, D.E., Walukiewicz, W., Ager, J.W. III, Haller, E.E., Chan, K.T., Tanigawa, S.: Annealing studies of low-temperature-grown GaAs:Be. J. Appl. Phys. 71, 1699–1707 (1992)
44. Chan, L.Y., Yu, K.M., Ben-Tzur, M., Haller, E.E., Jaklevic, J.M., Walukiewicz, W., Hanson, C.M.: Lattice location of diffused Zn atoms in GaAs and InP single crystals. J. Appl. Phys. 69, 2998–2300 (1991)
45. Wu, Y.-F., Keller, B.P., Keller, S., Kapolnek, D., Kozodoy, P., Denbaars, S.P., Mishra, U.K.: High power AlGaN/GaN HEMTs for microwave applications. Solid-State Electron. 41, 1569–1574 (1997)
46. Davydov, V.Y., Klochkin, A.A., Seisyan, R.P., Emtsev, V.V., Ivanov, S.V., Bechstedt, F., Furthmuller, J., Harima, H., Murdryi, A.V., Aderhold, J., Semchinova, O., Graul, J.: Absorption and emission of hexagonal InN. Evidence of narrow fundamental band gap. Phys. stat. sol. B 229, R1–R2 (2002)
47. Wu, J., Walukiewicz, W., Yu, K.M., Ager, J.W. III, Haller, E.E., Lu, H., Schaff, W.J., Saito, Y., Nanishi, Y.: Unusual properties of the fundamental band gap of InN. Appl. Phys. Lett. 80, 3967–3969 (2002)
48. Wu, J., Walukiewicz, W., Yu, K.M., Ager, J.W. III, Li, S.X., Haller, E.E., Lu, H., Schaff, W.J.: Universal bandgap bowing in group-III nitride alloys. Solid State Commun. 127, 411–414 (2003)
49. Sato, Y., Sato, S.: Growth of InN thin films by hydride vapor phase epitaxy. J. Cryst. Growth 144, 15–19 (1994)
50. Weyers, M., Sato, M., Ando, H.: Red shift of photoluminescence and absorption in dilute GaAsN alloy layers. Jpn. J. Appl. Phys. 31, L853–L855 (1992)
51. Uesugi, K., Marooka, N., Suemune, I.: Reexamination of N composition dependence of coherently grown GaNAs band gap energy with high-resolution X-ray diffraction mapping measurements. Appl. Phys. Lett. 74, 1254–1256 (1999)
52. Shan, W., Walukiewicz, W., Ager, J.W. III, Haller, E.E., Geisz, J.F., Friedman, D.J., Olson, J.M., Kurtz, S.R.: Band anticrossing in GaInNAs alloys. Phys. Rev. Lett. 82, 1221–1224 (1999)
53. Walukiewicz, W., Shan, W., Ager, J.W. III, Chamberlin, D.R., Haller, E.E., Geisz, J.F., Friedman, D.J., Olson, J.M., Kurtz, S.R.: Nitrogen-induced modification of the electronic structure of group III-N-V alloys. In: Proceedings of the 195th Meeting of the Electrochemical Society, vols. 99–11, p. 190 (1999)
54. Skierbiszewski, C., Perlin, P., Wisniewski, P., Knap, W., Suski, T., Walukiewicz, W., Shan, W., Yu, K.M., Ager, J.W., Haller, E.E., Geisz, J.F.,

Olson, J.M.: Large, nitrogen-induced increase of the electron effective mass in $In_yGa_{1-y}N_xAs_{1-x}$. Appl. Phys. Lett. 76, 2409–2411 (2000)

55. Yu, M., Walukiewicz, W., Shan, W., Ager, J.W. III, Wu, J., Haller, E.E., Geisz, J.F., Friedman, D.J., Olson, J.M.: Nitrogen-induced increase of the maximum electron concentration in group III-N-V alloys. Phys. Rev. B 61, R13337–R13340 (2000)

56. Walukiewicz, W., Shan, W., Yu, K.M., Ager, J.W. III, Haller, E.E., Miotkowski, I., Seong, M.J., Alawadhi, H., Ramdas, A.K.: Interaction of localized electronic states with the conduction band: Band anticrossing in II–VI semiconductor ternaries. Phys. Rev. Lett. 85, 1552–1555 (2000)

57. Lin, W., Guo, S.P., Tamargo, M.C., Kuskovski, I., Tian, C., Neumark, G.: Enhancement of p-type doping of ZnSe using a modified (N+Te) delta-doping technique. Appl. Phys. Lett. 76, 2205–2207 (2000)

58. Qiu, J., DePuydt, J.M., Cheng, H., Haase, M.A.: Heavily doped p-ZnSe:N grown by molecular beam epitaxy. Appl. Phys. Lett. 59, 2992–2994 (1991)

59. Chadi, D.J.: Doping in ZnSe, ZnTe, MgSe, and MgTe wide-band-gap semiconductors. Phys. Rev. Lett. 72, 534–537 (1994)

60. Garcia, A., Northrup, J.E.: Compensation of p-type doping in ZnSe: The role of impurity-native defect complexes. Phys. Rev. Lett. 74, 1131–1134 (1995)

61. Chen, A.L., Walukiewicz, W., Duxstad, K., Haller, E.E.: Migration of compensating defects in p-type ZnSe during annealing. Appl. Phys. Lett. 68, 1522–1524 (1996)

62. Walukiewicz, W.: Activation of shallow dopants in II–VI compounds. J. Cryst. Growth 159, 244–247 (1996)

63. Zhang, S.B., Wei, S.-H., Zunger, A.: A phenomenological model for systematization and prediction of doping limits in II–VI and I–III–VI$_2$ compounds. J. Appl. Phys. 83, 3192–3196 (1998)

64. Ferreira, S.O., Sitter, H., Faschinger, W.: Molecular beam epitaxial doping of ZnMgSe using ZnCl$_2$. Appl. Phys. Lett. 66, 1518–1520 (1995)

65. Ferreira, S.O., Sitter, H., Faschinger, W., Brunthaler, G.: Growth of highly doped p-type ZnTe layers on GaAs using a nitrogen DC plasma cell. J. Cryst. Growth 140, 282–286 (1994)

66. Ogawa, H., Ifran, G.S., Nakayama, H., Nishio, M., Yoshida, A.: Growth of low-resistivity n-type ZnTe by metalorganic vapor phase epitaxy. Jpn. J. Appl. Phys. 33, L980–L982 (1994)

67. Ohtsuka, T., Kawamata, J., Zhu, Z., Yao, T.: P-type CdSe grown by molecular beam epitaxy using a nitrogen plasma source. Appl. Phys. Lett. 65, 466–468 (1994)

68. Sondergeld, M.: Two-photon absorption by envelope-hole coupled exciton states in cubic ZnSe. I. Energy spectrum of the H$_d$ term and valence-band parameters. Phys. Stat. Solidi, B 81, 253–262 (1977)

69. Contreras, M.A., Egaas, B., Ramanathan, K., Hiltner, J., Swartzlander, A., Hasoon, F., Noufi, R.: Progress toward 20% efficiency in Cu(In,Ga)Se$_2$ polycrystalline thin-film solar cells. Prog. Photovolt. Res. Appl. 7, 311–316 (1999)

70. Jasenek, A., Rau, U.: Defect generation in Cu(In,Ga)Se$_2$ heterojunction solar cells by high-energy electron and proton irradiation. J. Appl. Phys. 90, 650–658 (2001)

71. Turcu, M., Kotschau, I.M., Rau, U.: Composition dependence of defect energies and band alignments in the Cu(In$_{1-x}$Ga$_x$)(Se$_{1-y}$S$_y$)$_2$ alloy system. J. Appl. Phys. 91, 1391–1397 (2002)

72. Zhang, S.B., Wei, S.-H., Zunger, A.: A phenomenological model for systematiza-
    tion and prediction of doping limits in Il–VI and I–III–VI$_2$ compounds. J. Appl.
    Phys. 83, 3192–3196 (1998)
73. Siebentritt, S.: Wide gap chalcopyrites: material properties and solar cells. Thin
    Solid Films 403–404, 1–8 (2002)

Confine Cu to Increase Cu-Chalcopyrite Solar Cell Voltage

J.A. Van Vechten

Given that Cu leaks out of Cu-chalcopyrite crystals when and where the Fermi level is 0.8 eV above the valence band edge and that this leakage causes the defect species known as N_1 and N_2, one may conclude that to increase solar cell operating voltages above about 0.8 eV it will be necessary to stop this leak and to confine the Cu to the absorber grains. In this chapter, a simple explanation is given as to why leakage of Cu produces anion vacancies that have been linked to N_1, and complexes consisting of group VI atoms on group III sites with Cu atoms on group III sites and Cu vacancies, which are proposed to be the N_2 defects. It is also argued that it will be practical to confine the Cu with amorphous SiON or amorphous TiN between grains and between the absorber and the n-type layer and with a dense random packing of hard-spheres-type amorphous metal as the p-type contact.

4.1 Introduction

Cu-chalcopyrite compounds and alloys have properties that make them interesting for the production of solar cells, particularly for use in space satellites [1–12]. (They are also interesting for the production of solid-state light-emitting devices (LEDs), particularly because they present lesser piezoelectric problems than do the group III nitrides that currently dominate that industry [4].) They are direct-gap semiconductors with gaps varying from 1.0 eV (CuInSe$_2$) to 3.5 eV (CuAlS$_2$) so that they span the AM0 solar spectrum almost completely. They can be produced as thin polycrystalline films on inexpensive flexible or rigid substrates by a wide variety of potentially inexpensive processes, including nonvacuum printing [11]. They are remarkably resistant to degradation due to high-energy electron and proton irradiation as is encountered on satellites in space; they are equaled in this regard only by rigid InP-based solar cells. For use on satellites weight and stowed volume are critical figures of merit. Even if they achieve 35% AM0 efficiency at end-of-life, the rigid cells are limited to less than 110 W/kg; whereas if only 10% AM0 efficiency is achieved at end-of-life, thin film cells on flexible substrates can provide more than 160 W/kg [12]. The advantage of thin flexible film restowed volume is probably even greater.

Solar cells with absorber layer compositions between $CuInSe_2$ and $CuGa_{0.3}$ $In_{0.7}Se_2$ are remarkably efficient with the long-wavelength portion of the solar spectrum, i.e., with photons having less than 1.2 eV energy. However, attempts to make good use of higher photon energies by employing absorber materials with larger direct bandgaps and by obtaining higher operating voltages have been frustrated for more than a decade. For terrestrial use, where stowed volume and power per unit weight are not as important as are balance of system expenses, which include mounting the cells, protecting them from the environment, collecting and conditioning the current, and bridging around series-connected cells that are temporarily in shadows, all of which are strongly correlated to AM1.5 efficiency and maximum power point voltage, the commercial prospects for Cu-chalcopyrite solar cells are dependent upon the finding of practical ways to increase efficiency and operating voltages.

In this author's view, the root cause of the problems that are frustrating attempts to increase operating voltages and to make good use of absorbers with larger bandgaps is already well established. When and where the Fermi level, E_F, is more than 0.8 eV above the valence band edge, E_V, i.e., when and where $E_F - E_V > 0.8$ eV, Cu emerges from grains of Cu-chalcopyrite materials [1, 2]. This loss of Cu has been correlated with the generation of two types of defects that are known as N_1 and N_2 and that strongly degrade the performance of the cell. The N_1 defects display donor levels well above midgap and have been correlated to group VI vacancies, $\square_{VI}$, that may or may not be bunched near the surfaces of absorber grains [1,2,13,14] (Chap.5). This author sees no reason to doubt the identification of the N_1 defects with anion vacancies or anion vacancy complexes as suggested by Niki et al. [14]. Note that Niki et al. used positron annihilation lifetime measurements in their studies; this method has the virtue that it sees only vacancies and other free space and is not confused by atom-derived defects. However, this author has not seen an explanation why loss of Cu should produce anion vacancies and so will provide one here.

The N_2 defects display an acceptor level that is 0.3 eV above E_V throughout the $Cu(InGa)Se_2$ alloy range and rises with substitution of S for Se to 0.5 eV above E_V in $Cu(InGa)S_2$ [15]. These defects are often the most important bulk recombination centers in solar absorber grains. It appears that they also have a shallower acceptor level that is 0.3 eV above E_V in $CuGaS_2$ and that pins E_F at metal contacts to the p-type material [16]. The fact that the higher acceptor level trends higher into the bandgap as one alloys S for Se, and thus increases the ionicity implies that this level is atom derived. Vacancy-derived ionization levels trend shallower with increasing ionicity while atom-derived levels trend deeper because of the fundamental difference between the two types of energy levels [17]. In its neutral state, an atom-derived acceptor level binds a hole to an atom-core potential while a vacancy-derived acceptor level simply has a neutral free space. In its ionized state, an atom-derived acceptor level has released the hole to the valance band distribution and a vacancy-derived acceptor level has attracted negative charge to a distribution on the atoms around the cavity. For an atom-derived acceptor, the greater the ionicity and smaller the dielectric constant, the tighter is the hole bound. For a vacancy-derived acceptor, the more electropositive the atom that is missing, the easier it is for the surrounding atoms to attract negative charge. For donor levels the argument is the same with the sign of charge reversed. Thus, completely covalent Si has two donor and two acceptor levels all much deeper than midgap, whereas in Cu-chalcopyrites the Cu is so electropositive that its acceptor level is only $0.03\,\mathrm{eV}$ above E_V, which means that it is shallower than atom-derived acceptor levels. Thus, it is practical to dope Cu-chalcopyrite solar absorber materials p-type simply with Cu vacancies. The Se vacancies provide donor levels that are somewhat deeper than the Cu vacancies, the S vacancies being less so. The Ga and In vacancies provide acceptor levels that are deeper yet.

One can also distinguish atom-derived ionization levels from vacancy-derived ionization levels according to the much larger temperature dependence of the ionization (free) energies of the vacancy-derived levels [17].

The fact that the N_2 acceptor level is so insensitive to group III doping strongly suggests that the responsible atom is on the group III site and that the group III atoms are not part of any complex. If it is also true that the defect is intrinsic, i.e., not involving H or Na or some other impurity, then one is forced to the conclusion that the responsible atom is Cu on the group III site, Cu_{In}. However, this author has not seen a statement explaining why the outflow of Cu should lead to Cu antisite defects. Therefore, he will provide one here.

Even if one does not understand exactly what N_1 and N_2 are and exactly why the outflow of Cu produces them, one might conclude that in order to improve the efficiency and operating voltages of wider bandgap Cu-chalcopyrite solar cells, it will be necessary to stop the Cu from leaking out after the absorber layer has first been made p-type. It is argued here that it should be possible to do this with atomic diffusion barriers consisting of amorphous thin films of SiON or TiN or a four-component amorphous alloy thereof, which are insulating, between the grains and between the p-type and the n-type layers to form an MIS-type solar cell [18,19], or actually an SIS-type solar cell and with a very dense and stable amorphous metallic film of the type known as dense random packing of hard spheres (DRPHS)-type amorphous metal [20–24] for the contact to the p-type layer. It is predicted here that the DRPHS-type amorphous metal, in addition to being an extremely effective diffusion stop for Cu, will provide a very large work function, an absence of interface states, low surface recombination velocity and very high conductivity. It may be superior to the provision of a wide bandgap, transparent p-type semiconducting material, such as $BaCu_2S_2$ as the back-contact [25], which has been proposed as an alternative method to prevent Cu from leaking out of the back-contact by providing a source for Cu there.

4.2 Why a Cu Leak Causes Anion Vacancies and Cu Antisite Defects

It seems that many workers in this field have interpreted the theoretical paper by Zhang et al. [5] to state and to show that $CuInSe_2$ in particular and all Cu-chalcopyrites in general, can withstand a loss of Cu of more than 1% without ruining device properties because the material simply generates alloys between the Cu-chalcopyrite and related ordered defect compounds (ODCs) such as $\square_2Cu_3In_7Se_{12}$, $\square_2Cu_2In_3Se_{10}$, $\square_2Cu_1In_5Se_8$, and $\square_2In_4Se_6$. Actually, this is not true and the paper does not make any such claim. The inferred reactions, such as

$$4(CuInSe_2) \rightarrow \square_2Cu_1In_5Se_8 + 3Cu_i \uparrow \qquad (4.1)$$

where $Cu_i \uparrow$ denotes Cu interstitial atoms leaving the crystal, *does not conserve In atoms*. There are four In atoms on the left-hand side and five In atoms on the right-hand side. It would necessarily generate vacancies on the In sites, $\square_{In}$.

What Zhang et al. [5] actually do (correctly) is to describe the alloying of $CuInSe_2$ with the ODCs during crystal growth, where sources and sinks for In as well as for Cu and Se are present. This is fundamentally different from the situation of a solar cell under construction when the n-type CdS layer is being deposited upon the p-type Cu-chalcopyrite absorber layer so that bands are bent and E_F rises more than 0.8 eV above E_V within the Cu-chalcopyrite or subsequently in the life of the solar cell, where no source of In atoms needed to balance Eq. 4.1 is available.

It is more wrong to have the notion that the loss of Cu atoms can be accommodated within the absorber layer by the generation of ODCs than the failure of Eq. 4.1 to conserve In atoms. The loss of Cu atoms implies a loss of valence electron density. All sp^3 hybridized tetrahedral crystal structures, including the chalcopyrite and all of the ODC structures, require an average of four valence electrons per lattice site (counting the vacant sites in the ODC structures). One should distinguish "deviations from stoichiometry", where the actual valence electron density deviates from 4.0 per lattice site, from the situation when a chalcopyrite alloys with an ODC but the valence electron density remains ideal at 4.0. While the ideal ODCs are all stable, major deviations from stoichiometry would destabilize the material.

Furthermore, there is a limit to how much reactions like

$$0 \rightarrow \square_{Cu} + Cu_i \uparrow \qquad (4.2)$$

where 0 denotes perfect crystal, that do not preserve the crystal structure can run [26]. A crystal can deviate from its ideal structure, i.e., having sites for each constituent in the ideal proportion only on grain boundaries and to a limited extent in dislocation cores and twins. The magnitude of the limit is not more than 0.1% [26, 27]. The loss of Cu from Cu-chalcopyrite solar absorber layer where $E_F - E_V > 0.8\,eV$ far exceeds this limit. To stabilize the material as Cu atoms are lost in massive quantities, the solar absorber grains must reduce the number of their lattice sites, so that the valence electron density approaches four per lattice site. To preserve the crystal structure, they must do it by eliminating substantially equal numbers of group III sites as Cu sites and twice as many group VI sites as Cu sites. (See Eq. 10 in [27]).

Consider a Gedanken experiment, in which we remove 800 Cu atoms from $CuGaSe_2$ so that we thereby remove 800 valence electrons and then we restore the valence electron density by removing 200 lattice sites by annihilating 50 Cu vacancies, 50 Ga vacancies and 100 Se vacancies at a surface of the crystal. Where would the corresponding Ga and Se atoms go? One might expect that they would form elemental precipitates of Ga and Se corresponding to the balls of Ga and thin films of As that form on the surfaces of GaAs that has gone off stoichiometry on the Ga-rich side [27, 28], as well as forming some ODC, but this is not observed. Therefore, we initially place these Ga and Se atoms onto vacant Cu sites. Thus, we have

$$0 \rightarrow 800\square_{Cu} + 800Cu_i \uparrow \qquad (4.3)$$

$$\rightarrow 50(\square_{Cu} + \square_{Ga} + 2\square_{Se}) \uparrow +100Se_{Cu} + 50Ga_{Cu} + 600\square_{Cu} \qquad (4.4)$$

Here, we see the generation of anion vacancies (and complexes) that produce the N_1 defects until they annihilate. Now, we continue the Gedanken experiment by letting the S_{Cu} displace Ga atoms and then combine Ga antisites and Cu vacancies to form some ODC

$$\rightarrow 100Se_{Ga} + 150Ga_{Cu} + 500\square_{Cu} \qquad (4.5)$$

$$\rightarrow 100Se_{Ga} + 150(Ga_{Cu} + 2\square_{Cu}) \downarrow +200\square_{Cu} \qquad (4.6)$$
$$\text{ODC}$$

where the ODC forms as an alloy within the Cu-chalcopyrite absorber grain. We may further reduce the concentration of Cu vacancies by allowing Ga atoms to take vacant Cu sites and forming more ODCs, but this can only be done at the cost of forming Ga vacancies

$$\rightarrow 100Se_{Ga} + 50(Ga_{Cu} + 2\square_{Cu}) \downarrow +50\square Cu + 100\square_{Cu} \qquad (4.7)$$
$$\text{ODC}$$

At this point, we should consider the relative virtual energies of formation [26, 31] of Ga vacancies and Cu vacancies. Neumann [29] proposed on the basis of his consideration of the macroscopic cavity model for vacancies [30–32] that the virtual enthalpy of neutral Cu vacancy formation is 2.6 eV, while that of neutral Ga vacancy formation is 2.7 eV. Zhang et al. [5] argue that the enthalpy of formation of Cu vacancies is much less than 2.6 eV, perhaps 0.6 eV. One might combine the critical value for $E_F - E_V = 0.8$ eV with the ionization energy for the Cu vacancy (0.03 eV) to conclude an empirical value of $\Delta H_f(\square_{Cu}) = 0.77$ eV. With the macroscopic cavity model, one calculates the surface energy of a bubble the size and shape of the missing atom or atoms for the theoretical jellium metal having the same valence electron density as the semiconductor of interest and then adds the energies of the broken bonds relative for the semiconductor relative to the jellium metal. This will clearly give an upper boundary for the vacancy formation enthalpy because it does not allow for any relaxation of the semiconductor around the vacancy. For Si, Ge, GaAs, InP, etc., it seems that the magnitude of the overestimate inherent in the use of the macroscopic cavity model is of order 10%. For Cu vacancies in Cu-chalcopyrites, the overestimate may be substantially greater because the surrounding atoms move into the cavity almost 10%, whereas for vacancies in Si the surrounding atoms move out slightly, and because the lattice constants for the ODCs are somewhat smaller than for the corresponding Cu-chalcopyrites. This should be expected from copper's high lying d-core levels, which account for the red color of the metal, the d-like states at the top of the valence band, and the anomalously small values of the direct bandgap of Cu-chalcopyrites relative to Ag-chalcopyrites [5]. There is a severe manifestation of what this author calls "d-core dehybridization" [17] to reduce the strength of sp^3 tetrahedral bonding occurring in Cu-chalcopyrites. It appears that Neumann assumed a rather large value for the enthalpy of the broken bonds, relative to jellium, on the Cu vacancy where he should have assumed almost zero enthalpy and should have concluded that $\Delta H_f(\square_{Cu}) < 1.8$ eV. However, the overestimate here may not be very large. Anyhow, this would be a non sequitur; the law of mass action

$$[\square_{Cu}][Cu_i] = K \exp(-\Delta H/kT) \qquad (4.8)$$

where [ ] denote concentration and K is a constant, will drive reaction 4.2 to the right-hand side no matter how large the enthalpy of the reaction (ΔH) may be as long as Cu$_i$ leaks out of the crystal, so that its concentration does not rise.

We can be sure that the (virtual) enthalpy of formation of Ga$_{Cu}$ is small because Cu-chalcopyrites will disorder to a zinc-blende or delta phase at moderate temperatures. Given this and assuming that the enthalpy cost of Cu vacancies is not drastically less than that of Ga vacancies, we conclude that reaction 4.7 will proceed to minimize the free energy of the Cu-chalcopyrite grains in the absorber layer of the solar cell structure by generating some concentration of Ga vacancies so that it can render more Cu vacancies and Ga antisites into the ODC alloy. The Ga vacancies can morph as

$$\square_{Ga} \leftrightarrow Cu_{Ga} + \square_{Cu} \qquad (4.9)$$

under the influence of the local Fermi level and the ionization states of the defects. All three of these are acceptors and will certainly form complexes with Se_{Ga}antisite multiple donors, which will favor the right-hand side of reaction (4.9). Thus, we can conclude our Gedanken experiment with the formation of the second nearest-neighbor complexes

$$\rightarrow Se_{Ga}{}^{+3/+2/+1} Cu_{Ga}{}^{-2/-} \square_{Cu}{}^{-} \qquad (4.10)$$

at a concentration where a few percent of the initial number of Cu atoms leak out. In fact, the observed concentrations of N_2 is rather less than a percent of the leaked Cu atoms but it is easy to reconcile with this discrepancy, with a proposed formation of other possible complexes. Alternatively, one can adjust the ratio of complexes formed to Cu atoms emitted by adjusting the assumed ratio of the virtual enthalpy of Ga vacancy formation to the virtual enthalpy of Cu vacancy formation. It is the group III vacancies that largely determine the N_2 concentration because there are fewer of these than there are Se_{III} and $\square_{Cu}$ available to form the complex.

4.3 Is $Se_{Ga}{}^{+3/+2/+1} Cu_{Ga}{}^{-2/-} \square_{Cu}{}^{-} = N_2$?

Consider seven points of identification for the N_2 defects:

1. They are very effective recombination centers. The nominee clearly meets this point; it binds both electrons and holes in close proximity and has multiple ionization levels spread through the gap to allow for rapid radiative and nonradiative recombination processes.
2. They have a broad acceptor level. The Cu_{Ga} provides two acceptor levels at the appropriate depth. They are broadened by interaction with the other two ionized defects at a thermally determined distribution of spacings.
3. They are insensitive to group III alloying. The nominee complex does not involve a group III atom and because the defect that provides the acceptor level in question is on a group III site, it is maximally separated from the group III atoms that are present.
4. The acceptor level trends deeper with increasing ionicity as S is substituted for Se. As noted above, this implies that the level is atom-derived. The nominee has the Cu_{Ga} providing this acceptor level.
5. The concentration in best available solar cell material is minimal for about 30% Ga in $Cu(InGa)Se_2$. The increase with Ga content greater than 30% can be attributed to a larger volume of material, where $E_F - E_V > 0.8\,eV$. The decrease as Ga content increases from 0 to 30% can be attributed to the fact that it is harder to put Se and Cu onto Ga sites than onto In sites.

6. High-energy electron and proton irradiation introduces new N_2 at a rate that is only 1% and 2%, respectively, of primary displacements [6]. Consider the effect of displacing the various types of atoms. Displacing Cu atoms will have no effect because the Cu interstitial will simply occupy one of the many Cu vacancies already present, leaving the concentration of Cu vacancies unaffected. Displacing Se atoms produces a Se vacancy – Se interstitial pair that might recombine with no effect or might form a III_{Se}–Se_{III} antistructure pair, which would have little effect. Displacing a group III atom produces III_{Cu} and group III vacancies. These may recombine with no effect or as described above, the group III vacancy can, with lesser probability, form a new N_2 defect. Formation of a Se_{Cu} antisite is improbable.[1]

7. Irradiation-induced additional N_2 defects will thermally anneal out with an activation energy of 1.05 eV [6]. The nominee defects can be annealed out by Ga atomic migration. Jasenek showed that the observed activation energy is reasonable for such migration [6].

4.4 Is the High-Voltage Cu-Chalcopyrite Case Hopeless?

The loss of Cu reactions that ruin solar cell performance occur spontaneously where and when $E_F - E_V > 0.8$ eV, due to the band bending required to form a solar cell. This must be because they are lowering the free energy of the material under that condition and because no activation barrier is restraining them. They are examples of exothermic self-compensation reactions. However, the fact that they are reducing the free energy of the material does not in itself make the situation hopeless. It is well known that the reaction converting diamond to graphite is exothermic at standard temperature and pressure but "Diamonds are Forever" because the activation barrier is high. Attempts to make AlGaN alloys p-type were frustrated for 25 years by self-compensation reactions in which N vacancies were formed and ionized as donors by holes rising from any Fermi level that began to approach E_V, while N left the crystal:

$$h^+ \rightarrow \square_N^+ + N_g \uparrow \qquad (4.11)$$

---

[1]If the Se_i finds a Cu vacancy ($\square_{Cu}$), then we have $\square_{Se}^+ + Se_{Cu}^{+3}$, which is rather unstable because the enthalpy of formation of the Se_{Cu} is large because the difference in valence between Se and Cu is so large, five as compared to a difference in valence of just three between Se and III and also because of coulomb repulsion. The crystal can react by nearest-neighbor hopping of a III atom into the $\square_{Se}^+$ with little activation energy to form $III_{Se}^{-2} + \square_{III}^- + Se_{Cu}^{+3}$, which will then react by second nearest-neighbor hopping of the Se antisite atom to $III_{Se}^{-2} + Se_{III}^{+2} + \square_{Cu}^-$ and the $\square_{Cu}^-$ will just replace the one that was caught by the initial Se interstitial. (The charge state assignments assume that the Fermi level is near midgap, which is the usual situation during irradiation because the fast charged particles produce a large density of electron–hole pairs.)

Then, this frustration was suddenly relieved [33, 34] by a stratagem [35] of growing the alloy saturated with H (by molecular organic chemical vapor deposition, MOCVD or plasma enhanced CVD), which is an amphoteric dopant and keeps E_F near the intrinsic level (E_i), despite the deliberate incorporation of a large concentration of shallow acceptors so that the self-compensation reaction does not generate N vacancies. Then, after growth, the crystal is cooled to standard temperature and the H is removed by a nonthermal method so that the crystal becomes highly conductive p-type but reaction 4.11 does not proceed because it has too large an activation energy. Then the crystal is encapsulated to prevent H from getting back in from the environment, in particular from the humidity in ambient air, and to prevent N gas from getting out. Very efficient and powerful LEDs and other devices are now being made from the group III nitrides using this process. The critical factor controlling their lifetime is the encapsulation (Müller, G.O., private communication). A similar stratagem is now working with other wide bandgap semiconductors.

On the other hand, some self-compensation situations are hopeless. As hypothesized by Chadi and Chang [36], the prospect of getting good n-type conductivity in AlGaAs alloys with Al proportion greater than 22% is hopeless because shallow donors are compensated by an effect known as the "DX effect." According to Chadi and Chang, this DX effect simply involves a displacement a short distance off its lattice site of whatever shallow donor impurity that one has employed trying to grow n-type material. This short displacement converts the shallow donor into an acceptor so that each donor impurity that undergoes this displacement subtracts two electrons from the conduction band density of electrons. Because there is no X defect to eliminate, no long-range diffusion to inhibit, no atoms to encapsulate and no substantial activation barrier, according to this hypothesis, there is no hope of obtaining good n-type conductivity in wide-bandgap AlGaAs alloys.

For the present case of Cu-chalcopyrite solar cells the challenge is to stop the generation of N_1 and N_2 and the means proposed here is to do this by preventing the leakage of Cu from regions of the absorber grains, where $E_F - E_V > 0.8\,\text{eV}$. This author thinks that it may be practical to do this because Cu should be easier to confine than is N gas and because, if the concentration of Cu interstitials within the Cu-chalcopyrite absorber grains increases, the law of mass action relation 4.8 implies that the generation of more Cu vacancies will stop even if the ΔH is small. The problem seems, in principle, less difficult than that for the group III nitrides. However, cost constraints are more severe for terrestrial solar cells than for LEDs and lasers.

4.5 How to Confine Copper

To make an atomic diffusion barrier, one has to avoid grain boundaries and dislocation cores because atomic diffusion is rapid on these. This generally means that one should consider amorphous materials. One may also note that

interface state densities and surface recombination velocities at interfaces between semiconductors and nonconducting amorphous materials are generally
less than for semiconductor interfaces to non-pseudomorphic crystalline materials. The interface between (100)-Si and its amorphous thermal oxide is the
champion for both figures of merit. Therefore, this author will consider only
amorphous materials to propose mechanisms for improving Cu-chalcopyrite
solar cells.

Also, to provide an effective atomic diffusion barrier, one needs a material with as little free space between the atoms as possible. Consideration of
the space-filling capacity of hard spheres shows one of the advantages of having spheres of different sizes. For a single size of sphere, the densest packing
possible (face centered cubic, FCC, or hexagonal close-packed, HCP) leaves
about one-fourth of the space empty, but smaller spheres sized to fit into the
interstices can reduce the empty space. If one continues to add appropriately
sized smaller spheres, as is done in powder metallurgy technology, one can
approach perfect filling of space.

In the Si-IC industry amorphous films of insulating SiO_2, Si_3N_4, and TiN
are used. The amorphous alloy SiON is known to be a better diffusion stop
than either amorphous compound. TiN is also reputed to be a good diffusion
stop for Cu (Wagner, J.F., private communication). Therefore, this author
suggests the use of SiON, TiN or SiTiON amorphous thin films between Cu-
chalcopyrite absorber grains and between these p-type absorber grains and the
n-type material of the solar cell. The latter implies the formation of an SIS-
type solar cell [18, 19]; Card and Yang determined the ideal thickness of the
insulating later for an Si-MIS solar cell to be 2.0 nm. This author proposes
that the same thickness be used for Cu chalcopyrites. It would then seem
possible to eliminate the CdS buffer layer needed for present art cells and to
go directly to the n+ ZnO:Al layer. This would avoid the environmental and
health problems caused by Cd.

While the encapsulating layer between the p-type absorber and the n-
type material must allow sunlight to pass, there is no such constraint on the
back-contact. The use of a wider bandgap, lower index of refraction, p-type
semiconductor like $BaCu_2S_2$ [25] might offer a benefit of optical confinement
as well as a source of Cu that might reduce the outflow of Cu from the absorber
grains, but this author proposes that a DRPHS-type amorphous metal [20–
24] would serve a solar cell better by providing lower series resistance, lower
interface state density, lower surface recombination, better copper confinement
and good optical reflectivity. The work function may also be larger.

DRPHS-type amorphous metal films are formed simply by mixing together, usually by sputtering onto an unheated substrate, atoms of different
sizes. The range of atomic radii available does not allow one to achieve the
ideal of using a second size that fits ideally into the interstices of a mixture of
FCC and HCP stacking layers but a difference of 20% is sufficient to produce
amorphous material that is stable, not because of a disordered net of directed
covalent bonds as is present in a-SiO_2 and other glasses, but because the atoms

are behaving essentially as hard spheres locked by dense space filling. They are usually made with rare earth elements providing the largest spheres and transition metals providing the second largest. Greater stability is obtained when a third sized metal is employed. During the sputtering process the sputtering gas, usually Ar, is generally incorporated into the amorphous film in proportions that may reach 50 at.% [23, 24]. Thus, if one employs two metals in the sputtering target, one gets three different sized atoms in the DRPHS amorphous film. If one employs three metals in the sputtering target, one gets a four-component DRPHS amorphous film. These films are much more stable in their amorphous phase than are amorphous metal phases obtained by violent quenching because, in order to crystallize, they must separate the atoms of different sizes sufficiently so that the polycrystalline grains that would form are larger than the critical size for which the lowering of bulk energy would exceed the interface energy concomitant with formation of the polycrystalline phase. The material is the closest physical approximation to the theoretical jellium material that has been produced and there is very little thermodynamic driving force to promote the separation of different sized atoms. However, when phase separation begins, at temperatures of order 1000°C, the material usually explodes. This is because the sputtering gas atoms that were dissolved at such high concentration diffuse to the grain boundaries and convert to gas at extreme pressure.

DRPHS amorphous metals were originally interesting to theorists considering amorphous and jellium states [20, 21]. Then they were used to make magnetic bubble memories that captured a share of the commercial market for electronic memory for a few years [22, 23]. The predominant composition was Gd, Co, Mo and Ar. Then they were adapted [24] to store radioactive Kr, which is a major byproduct of nuclear power reactors, has a 10-year half-life that is about the worst possible value and is difficult to store by other methods. The decay product of Kr is Rb, which is highly corrosive to the metals one might use for high-pressure tanks to store the gas. Rb resides in the DRPHS amorphous metal as happily as does Kr and the recoil of the decay serves to stir up the amorphous structure and to reverse any phase separation that has begun. Kr diffuses rapidly through known polycrystalline materials and does not react chemically to a stable compound, except for its participation in the jellium binding of the DRPHS amorphous metal. The diffusivity of Kr in DRPHS amorphous metal is too small to measure and is calculated to be extremely small [24].

One should understand that Gd was used to make magnetic bubble memories because of its electronic properties, which were critical to those devices, and that Gd is expensive because it is very difficult to separate it from all the other rare earth elements that occur in mischmetal, the naturally occurring mixture of rare earth elements, which have very similar chemical properties. Mischmetal, which is denoted as Me (like the symbol of an element), costs only about $10/kg, and thus is practical for storing radioactive waste. The recommended composition [24], is 20 at.% Me, 60 at.% Fe, and 20 at.% Cr in the sputtering target. The fraction of Kr, which becomes Rb, in the film is about 25 at.%.

For use as a p-type contact to Cu-chalcopyrite solar absorber films this author would prefer to substitute Mo for Cr in the target composition used for Kr storage. This is because of the previous successful use of Mo with Cu-chalcopyrite solar cells and the larger work function of the element. The sputtering gas should be Ar or a mixture of Ar and He. Although not knowing of a measured value of the work function for the proposed DRPHS amorphous metal MeFeMoAr film, this author predicts it to be very large. This is because the theoretical value of the surface energy (tension) and the work function of jellium is an analytic function of electron density, which will be very large for the proposed film. Also, it was observed that DRPHS GdCoAr films absorbed H from the environment, which is a property of materials with functions enough to strip the electron from the proton and let the electron fall to the Fermi level while the proton adds to the jellium background.

This author predicts that the Cu-chalcopyrite to DRPHS interface will not generate interface states to pin the Fermi level, as is observed for polycrystalline metal interfaces [16]. This author predicts that the DRPHS material will act like a sponge for defects as it absorbs noble gas atoms during sputtering. A back-contact with a large work function and no interface pinning would improve the solar cell by increasing the width of the depletion layer.

References

1. Rau, U., Schock, H.W.: Cu(In,Ga)Se$_2$ solar cells. In: Archer, M.D., Hill, R. (eds.) Clean Electricity from Photovoltaics, pp. 277–296, Imperial College Press, London, (2001)
2. Siebentritt, S.: Wide gap chalcopyrites: material properties and solar cells. Thin Solid Films 403–404, 1–8 (2002)
3. Wagner, S., Shay, J.L., Migliorato, P., Kasper, H.M.: CuInSe$_2$/CdS heterojunction photovoltaic detectors. Appl. Phys. Lett. 25, 434–436 (1974)
4. Chichibu, S., Shirakata, S., Isomura, S., Nakanishi, H.: Visible and ultraviolet photoluminescence from Cu–III–VI2 chalcopyrite semiconductors grown by metalorganic vapor phase epitaxy. Jpn. J. Appl. Phys. 36, 1703–1714 (1997)
5. Zhang, S.B., Wei, S.H., Zunger, A., Katayama-Yoshida, H.: Defect physics of the CuInSe$_2$ chalcopyrite semiconductor. Phys. Rev. B 57, 9642–9656 (1998)
6. Jasenek, A.: Eigenschaften von Defekten in Cu(In,Ga)Se$_2$ nach Elektronen- und Protonenbestrahlung. Shaker Verlag, Aachen (2002).
7. Jasenek, A., Rau, U.: Defect generation in Cu(In,Ga)Se$_2$ heterojunction solar cells by high-energy electron and proton irradiation. J. Appl. Phys. 90, 650–658 (2001)
8. Jasenek, A., Schock, H.W., Werner, J.H., Rau, U.: Defect annealing in Cu(In,Ga)Se$_2$ heterojunction solar cells after high-energy electron irradiation. Appl. Phys. Lett. 79, 2922–2924 (2001)
9. Jasenek, A., Rau, U., Weinert, K., Schock, H.W., Werner, J.H.: Illumination-induced recovery of Cu(In,Ga)Se$_2$ solar cells after high-energy electron irradiation. Appl. Phys. Lett. 82, 1410–1412 (2003)

10. Kushiya, K.: Improvement of electrical yield in the fabrication of CIGS-based thin-film modules. Thin Solid Films 387, 257–261 (2001)
11. Kapur, V., Bansal, A., Le, P., Asensio, O.I.: Non-vacuum processing of $CuIn_{1-x}Ga_xSe_2$ solar cells on rigid and flexible substrates using nanoparticle precursor inks. Thin Solid Films 431–432, 53–57 (2003)
12. Senft, D.: Novel methods to improve the efficiency of copper–indium–gallium–diselenide solar cells AF04-25, http://www.acq.osd.mil/sadbul/sbir/solicitations/sbir041/word/af041.doc, 1 Oct. 2003.
13. Cahen, D., Noufi, R.: Defect chemical explanation for the effect of air anneal on $CdS/CuInSe_2$ solar cell performance. Appl. Phys. Lett. 54, 558–590 (1989)
14. Niki, S., Suzuki, R., Ishibashi, S., Ohdaira, T., Fons, P.J., Yamada, A., Oyanai, H., Wada, T., Kimura, R., Nakada, T.: Anion vacancies in $CuInSe_2$. Thin Solid Films 387, 129–134 (2001)
15. Turcu, M., Ktschau, I.M., Rau, U.: Composition dependence of defect energies and band alignments in the $Cu(In_{1-x}Ga_x)(Se_{1-y}S_y)_2$ alloy system. J. Appl. Phys. 91, 1391–1399 (2002)
16. Sugiyama, M., Nakai, R., Nakanishi, H., Chichibu, S.F.: Fermi-level pinning at the metal/p-type $CuGaS_2$ interfaces. J. Appl. Phys. 92, 7317–7319 (2002)
17. Van Vechten, J.A.: A simple mans view of the thermochemistry of semiconductors. In: Keller, S.P. (ed.) Handbook on Semiconductors, vol. 3, Materials, Properties, and Preparation, p. 15. North Holland, Amsterdam (1980)
18. Sze, S.M.: Physics of Semiconductor Devices, 2nd edn, p 824. John Wiley & Sons, New York (1981)
19. Card, H.C., Yang, E.S.: MIS-Schottky theory under conditions of optical carrier generation in solar cells. Appl. Phys. Lett. 29, 51–54 (1976)
20. Nowick, A.S., Mader, S.R.: A hard-sphere model to simulate alloy thin films. IBM J. Res. Develop. 9, 358–365 (1965)
21. Cargill, G.S. III: Dense random packing of hard spheres as structural model for noncrystalline metallic solids. J. Appl. Phys. 41, 2248–2253 (1970)
22. Chaudhari, P., Cuomo, J.J., Gambino, R.J.: Amorphous metallic films for bubble domain applications. IBM J. Res. Develop. 17, 66–68 (1973)
23. Cuomo, J.J., Gambino, R.J.: Incorporation of rare gases in sputtered amorphous metal films. J. Vac. Sci. Technol. 14, 152–157 (1977)
24. Van Vechten, J.A., Gambino, R.J., Cuomo, J.J.: Encapsulation of radioactive noble gas waste in amorphous alloy. IBM J. Res. Develop. 23, 278–285 (1979)
25. Park, S., Keszler, D.A., Valencia, M.M., Hoffman, R.L., Bender, J.P., Wager, J.F.: Transparent p-type conducting $BaCu_2S_2$ films. Appl. Phys. Lett. 80, 4393–4395 (2002)
26. Kröger, F.A.: Defect chemistry in crystalline solids. In: Huggins, R.A., Bube, R.H., Robert, R.W. (eds.) Annual Reviews of Material Science, vol. 7, pp 449–475. Annual Reviews Inc., Palo Alto (1977)
27. Van Vechten, J.A.: Simple theory of defects in ternary and multinary semiconductors or what is wrong with binaries and can we fix it by going to ternaries. In: Deb, S.K., Zunger, A. (eds.) Proceedings of the 7th International Conference on Ternary and Multinary Compounds, Snowmass, CO, USA, 1986, pp 423–427. MRS, Pittsburg (1987)
28. Freeouf, J.L., Woodall, J.M.: Schottky barriers: an effective work function model. Appl. Phys. Lett. 39, 727–729 (1981)
29. Neumann, H.: Vacancy formation enthalpies in $A^I B^{III} C_2^{VI}$ chalcopyrite semiconductors. Cryst. Res. Technol. 18, 901–906 (1983)

30. Phillips, J.C., Van Vechten, J.A.: Macroscopic model of formation of vacancies in semiconductors. Phys. Rev. Lett. 30, 220–223 (1973)
31. Van Vechten, J.A.: Simple theoretical estimates of the Schottky constants and virtual-enthalpies of single vacancy formation in zinc-blende and wurtzite type semiconductors. J. Electrochem. Soc. 122, 1746–1746 (1975)
32. Van Vechten, J.A.: Divacancy binding enthalpy in semiconductors. Phys. Rev. B 11, 3910–3917 (1975)
33. Amano, H., Kito, M., Hiramatsu, K., Akasaki, I.: P-type conduction in Mg-doped GaN treated with low-energy electron beam irradiation (LEEBI). Jpn. J. Appl. Phys. 28, L2112–L2114 (1989)
34. Nakamura, S., Senoh, M., Mukai, T.: Highly P-typed Mg-doped GaN films grown with GaN buffer layers. Jpn. J. Appl. Phys. 30, L1708–L1710 (1991)
35. Van Vechten, J.A., Zook, J.D., Horning, R.D., Goldenberg, B.: Defeating compensation in wide gap semiconductors by growing in H that is removed by low temperature de-ionizing radiation. Jpn. J. Appl. Phys. 31, 3662–3663 (1991)
36. Chadi, D.J., Chang, K.J.: Energetics of DX-center formation in GaAs and $Al_xGa_{1-x}As$ alloys. Phys. Rev. B 39, 10063–10074 (1989)

5

Photocapacitance Spectroscopy in Copper Indium Diselenide Alloys

J.D. Cohen, J.T. Heath, and W.N. Shafarman

5.1 Introduction

Research into the electronic properties of copper-based chalcopyrites, particularly $CuIn_{1-x}Ga_xSe_2$ (CIGS), continues to be very active because of the success these materials have had toward the development of a promising thin film solar cell technology. However, the detailed relationship between the electronic properties of these materials and their associated photovoltaic device performance remains ambiguous. For example, while efficiencies for such devices in the laboratory in some cases are coming close to their maximum theoretical values [1], much lower efficiencies are generally obtained when using fabrication methods more suitable for large-scale manufacturing. The reasons for this are not at all clear. Therefore, there is a critical need for new insights into the properties of CIGS alloys that could help account for the variations in the performance of these devices.

During the 1990s, the primary methods employed to study the electronic properties of CIGS thin film materials have been admittance spectroscopy [2–5] and deep-level transient spectroscopy (DLTS) [3,6,7]. These techniques employ capacitance measurements of the semiconductor junction that lies at the heart of the photovoltaic device itself. This is a great advantage because it means that such techniques are able to deduce defect distributions within an actual working device. Both the admittance technique and DLTS deduce such defect distributions by monitoring the capacitance response of the junction due to the capture and thermal emission of trapped charge carriers. Therefore, these methods are most applicable to the study of majority carrier traps, which are defects generally lying in the lower half of the gap.

Recently, we demonstrated that an alternative technique, drive-level capacitance profiling (DLCP) [8], could be used to more reliably determine the spatial variation of such hole traps as distinct from their thermal energy distributions. In particular, it was then possible to clearly distinguish between defects localized near the barrier junction from those distributed more uniformly throughout the CIGS absorber [9,10]. This technique also allowed a more accurate determination of the free carrier (hole) densities within the undepleted regions of such photovoltaic devices. Nonetheless, this method is based upon the same physical principles as admittance spectroscopy and therefore can only be used to address majority carrier processes.

Evidence that the majority carrier traps detected by such methods actually affect overall device performance is mixed. A fairly strong correlation was found between the open-circuit voltage and a defect characterized by admittance spectroscopy (labeled "N2") at levels exceeding $10^{16}\,cm^2$ [11]. However, in those studies, to obtain quantitative agreement with the observations, an unusually large minority (electron) carrier cross section of $10^{-13}\,cm^2$ had to be assumed. It should also be noted that such large N2 defect levels are not generally observed in most of the CIGS materials produced by different laboratories. At the same time, attempts to correlate device performance with DLCP determined defect densities in devices from sources of CIGS samples in USA generally have not succeeded very well [10, Young, D.L., private communication]. Thus, the impact on devices of the majority carrier traps detected by the above methods is currently not clear.

In spite of this, we believe that defect levels detected using admittance spectroscopy and related methods provide one important component toward understanding the behavior of CIGS-based photovoltaic devices, particularly when such defects exceed concentrations of $10^{16}\,cm^{-3}$. However, the above examples suggest that there are additional aspects of electronic structure in these materials, not revealed by the mentioned methods, that are perhaps even more important for understanding the device performance. Specifically, it seems crucial to be able to examine defect levels closer to the conduction band that may interact strongly with the electron carriers in these materials. Such states will almost certainly impede minority carrier collection and could also dominate carrier recombination in these materials.

One method that can provide such information about defect transitions across the entire bandgap is that of sub-bandgap optical absorption. Such measurements can also reveal the degree of bandtailing from the band edges, which is an indicator of the disorder within the semiconductor [12, 13]. The width of the bandtail distribution, characterized by the Urbach energy (E_U) also provides insight into carrier properties since carrier trapping into and out of such bandtail states usually limits their effective mobility [14, 15]. Because such sub-bandgap absorption can reveal defect states not observable by standard admittance spectroscopic measurements, one may also be able to identify key defect levels associated with minority carrier trapping and/or recombination.

The earliest sub-bandgap absorption measurements on the copper-based chalcopyrite materials were carried out on relatively thick crystalline samples [16]. These kinds of studies revealed that the Urbach energy in high-quality single-crystal samples could be extremely narrow: less than 10 meV at room temperature and very temperature dependent since thermal disorder dominates in this case [17]. Unfortunately, such direct optical transmission/reflectivity measurements are not sensitive enough to examine deep defects at levels below 10^{17} cm^{-3} in thin film semiconductors. Because of this, two alternative sub-bandgap optical techniques had previously been developed for the study of amorphous silicon thin films with good results: photothermal deflection spectroscopy (PDS) [18] and the constant photocurrent method (CPM) [19]. Both these methods were successfully applied to CIGS thin film samples within the last couple of years [20]. However, such methods are unsuitable if one wants to characterize films incorporated into actual photovoltaic devices. This is because the additional materials used for contacting (CdS, ZnO, etc.) exhibit absorption of their own in the sub-bandgap energy regime that greatly exceeds that due to the deep defects in a 2 μm-thick CIGS absorber. Moreover, the ability to characterize CIGS films within their actual working device environment may be crucially important because of evidence that the barrier junction can actually significantly modify the electronic properties of chalcopyrite films over distances comparable with their thickness (for example, see [21]).

In contrast with the sub-bandgap optical methods mentioned above, the transient photocapacitance (TPC) spectroscopic technique can be applied directly to photovoltaic devices without further modification. Moreover, this method can be used in conjunction with the transient junction photocurrent measurements to identify whether defect transitions involve the majority (valence) or minority (conduction) band, and also provide some information about relative minority vs majority carrier collection efficiencies. The focus of this chapter will be to provide the details of how these techniques work and to discuss the kinds of information they have begun to provide us with about the defect structure in copper-based chalcopyrite materials, as well as the potential relevance of this information toward understanding the behavior of photovoltaic devices.

Historically, TPC spectroscopy was first developed for the study of amorphous silicon materials [22,23]. In studies of those materials, TPC, PDS, and CPM sub-bandgap optical measurements were all employed throughout the 1980s and the early 1990s to deduce the density of states across the mobility gap. The TPC method had the advantage that it could be applied to sandwich device structures similar to those being used for amorphous silicon photovoltaic devices. It had the disadvantage that because those materials are nearly intrinsic, elevated temperature and low frequencies had to be employed in order to clearly monitor the junction capacitance signal. For these reasons, we have found the TPC method to actually be much more ideally suited to the study of CIGS. The results obtained by the photocapacitance method applied to CIGS were first reported by us only a few years ago [24,25].

5.2 The Photocapacitance Method

The principle and implementation of the TPC method is similar to the familiar DLTS technique [26,27]. The semiconductor junction held under reverse bias is subjected periodically to a voltage "filling pulse" that allows majority carriers to move into the previously depleted region and be captured. Following these pulses a capacitance transient can be observed as holes are thermally emitted out of the majority carrier traps and leave the depletion region. In the TPC measurement, however, we introduce sub-bandgap monochromatic light to induce optical transitions in addition to any thermal ones. Actually, we apply light after *every other* filling pulse so that the capacitance transients with and without light present can be subtracted. A schematic of the time sequence of filling pulses, sub-bandgap light excitation, and resulting capacitance transients is shown in Fig. 5.1.

The capacitance transient signals are integrated over a "time window"— between t_1 and t_2—following each filling pulse (see Fig. 5.1), and the *difference* is obtained for sequential pairs of transients, with and without light present. This difference is normalized to the photon flux, and thus yields the *photocapacitance signal* (S_{TPC}) at the photon energy (E_{opt}) selected by the monochromator. Specifically:

$$S_{\mathrm{TPC}}(E_{\mathrm{opt}}) \equiv \frac{\int_{t_1}^{t_2} C_{\mathrm{light}}(t)\, dt - \int_{t_1}^{t_2} C_{\mathrm{dark}}(t)\, dt}{\text{Photon flux at } E_{\mathrm{opt}}}. \tag{5.1}$$

Repeating this over the full range of sub-bandgap photon energies available yields the *photocapacitance spectrum*. In such measurements, it is important to work in the linear regime (low enough light levels) such that S_{TPC} will be independent of the intensity of light used. An example of such a photocapacitance spectrum for an unalloyed $CuInSe_2$ sample device is shown in Fig. 5.2.

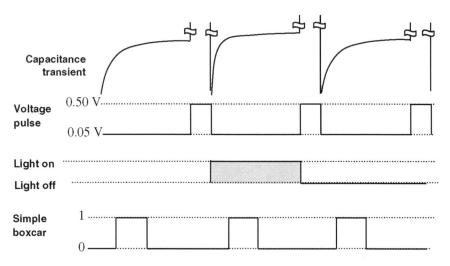

Fig. 5.1. Measurement timing for the transient photocapacitance measurement. The voltage pulse is applied on top of an applied DC bias, which is typically $-0.5\,\mathrm{V}$

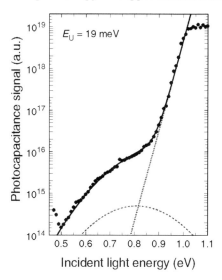

Fig. 5.2. Photocapacitance spectrum for a CuInSe$_2$ polycrystalline sample (D008). The data points have been fit assuming a Gaussian shaped defect band centered at 0.81 eV plus an exponential bandtail with a characteristic (Urbach) energy of 19 meV

The resulting spectrum is very similar to what we would obtain if we could carry out a true sub-bandgap optical absorption spectrum of the absorber layer, but also manifests some important differences. In general, optical absorption at E_{opt} is represented by the integral

$$P(E_{\mathrm{opt}}) = \int |\langle i\,|\mathrm{ex}|\,f\rangle|^2 g_{\mathrm{unocc}}(E)\, g_{\mathrm{occ}}(E - E_{\mathrm{opt}})\,\mathrm{d}E, \qquad (5.2)$$

where $|\langle i\,|\mathrm{ex}|\,f\rangle|$ represents the optical matrix element between occupied and unoccupied states, g_{occ} and g_{unocc}, separated by energy E_{opt}. In the case of a sub-bandgap spectrum involving defect states there will be two contributions to Eq. 5.2; first, transitions from the valence band to unoccupied gap states

$$P_1(E_{\mathrm{opt}}) = \int_{E_{\mathrm{F}}}^{E_{\mathrm{V}}+E_{\mathrm{opt}}} |\langle i\,|\mathrm{ex}|\,f\rangle|^2 g(E) g_{\mathrm{V}}(E - E_{\mathrm{opt}})\,\mathrm{d}E \qquad (5.3)$$

and second, transitions from occupied gap states to the conduction band

$$P_2(E_{\mathrm{opt}}) = \int_{E_{\mathrm{C}}-E_{\mathrm{opt}}}^{E_{\mathrm{F}}} |\langle i\,|\mathrm{ex}|\,f\rangle|^2 g(E) g_{\mathrm{C}}(E + E_{\mathrm{opt}})\,\mathrm{d}E. \qquad (5.4)$$

In both Eqs. 5.3 and 5.4, $g(E)$ is now being used to denote states lying specifically within the bandgap, while g_{V} and g_{C} denote states in the valence band

and conduction bands, respectively. Note that Eq. 5.3 is valid only if $E_V + E_{opt}$ lies higher in energy than E_F; otherwise P_1 is zero. Similarly, Eq. 5.4 is valid only when $E_C - E_{opt}$ lies below E_F; otherwise P_2 is zero.

In an optical absorption measurement, both integrals contribute to the observed attenuation of light with equal weight. In the TPC measurement, however, the signal results from the amount of charge released from the depletion region in response to the light. For the moment, if we consider only excitation of valence band electrons into empty gap states and assume that all holes left in the valence band escape the depletion region, then the TPC signal will be given by an integral identical to that in Eq. 5.3, except that the lower limit (E_F) should be replaced by the demarcation energy between filled and empty gap states appropriate to the depletion region. Specifically, if the junction is subjected to repeated filling pulses and the time window of observation extends between t_1 and t_2 after each one, then the states lying within an energy E_e above E_V will have thermally emitted their trapped holes, where

$$E_e = k_B T \log(\nu \tau). \tag{5.5}$$

Here ν denotes the thermal emission prefactor and τ is roughly equal to $(t_1 + t_2)/2$. Thus, for the TPC measurement, the lower limit of the integral in Eq. 5.3 should be replaced by $E_V + E_e$ as the lowest energy unoccupied gap state. (The upper limit in Eq. 5.4 is likewise replaced by $E_V + E_e$).

In the simplest analysis, the optical matrix element is assumed to be energy independent. The approximate density of gap states is then obtained by differentiating the TPC spectrum with respect to E_{opt}, as can be seen by examining Eq. 5.3 and assuming that the density of states in the valence band is slowly varying in energy compared to features in $g(E)$. The TPC data for a polycrystalline CuInSe$_2$ produced at IEC, displayed in Fig. 5.2, has been analyzed in precisely this fashion: integrating over the Gaussian-shaped defect band plus the exponential bandtail indicated provides a very good fit to the TPC spectrum as shown by the solid line. Thus, we infer the existence of a prominent band of empty gap states lying roughly 0.8 eV above the valence band edge with a Gaussian energy width parameter of 0.13 eV (implying a FWHM close to 0.3 eV).

5.3 Photocapacitance Spectra for the Cu(InGa)Se$_2$ Alloys

Polycrystalline CIGS devices were fabricated at the Institute of Energy Conversion (IEC), University of Delaware. These films were grown about 2 μm thick using four-source elemental evaporation [28, 29]. All of the IEC films were deposited on soda-lime glass that had been coated with a 1 μm Mo layer. To form devices, a chemical bath deposition was used to deposit 30–40 nm of CdS, then ZnO:Al was sputtered to form the top contact, with evaporated

Ni/Al grids. The CIGS film compositions, determined by energy dispersive X-ray spectroscopy, are slightly Cu poor, with $\mathrm{Cu}/(\mathrm{In} + \mathrm{Ga}) = 0.8$–$0.9$.

The CIGS films were prepared with Ga fractions ranging between 0 and 0.8 and are listed in Table 5.1. A majority of the CIGS samples studied had ratios of $\mathrm{Ga}/(\mathrm{In}+\mathrm{Ga}) = 0.3$, resulting in a bandgap, E_g, near 1.2 eV. Although most of the CIGS films were deposited at 550°C, a couple of CIGS samples studied were deposited at lower temperatures. Auger depth profiles indicated uniform Ga compositions in all cases.

Examples of TPC spectra for a range of $\mathrm{CuIn}_{1-x}\mathrm{Ga}_x\mathrm{Se}_2$ (CIGS) samples with varying Ga fractions are displayed in Fig. 5.3. One can clearly identify the increasing bandgap with the Ga content by noting the energy at which the exponential bandtail begins to flatten at higher optical energies. This occurs due to the attenuation of the light within the absorber once the optical gap is reached. That is, once the absorption length becomes shorter than roughly 1 μm, the light will no longer penetrate far enough to induce gap state transitions over the entire depletion region. However, the more noteworthy aspect of this series of spectra is their similarity. In particular, the Gaussian defect band of transitions appears essentially identical, independent of Ga fraction: they are all centered at 0.8 ± 0.05 eV, with an FWHM near 0.3 eV. In addition, we have observed the same defect in photocapacitance spectra for an epitaxial single-crystal CIGS sample deposited by Angus Rockett at the University of Illinois [25]. Thus, we concluded that the observed defect band

Table 5.1. CIGS devices studied, in order from lowest to highest Ga content together with their properties

sample #	T_SS (°C)	E_g(eV)	$\dfrac{\mathrm{Ga}}{\mathrm{Ga+In}}$	E_U (meV)	V_oc(V)	J_sc (mA/cm^2)	FF (%)	η(%)
D469	550	1.04	0.0	18	0.46	36.8	66.6	11.3
D008	550	1.04	0.0	19	0.41	34.8	65.6	9.5
D982	550	1.16	0.27	17	0.53	33.3	62.6	11.0
D362	550	1.16	0.27	16	0.53	32.9	63.8	11.2
D427	550	1.17	0.29	19	0.61	31.7	74.7	14.5
D934	550	1.17	0.30	21	0.63	31.5	76.7	15.2
D400	550	1.19	0.32	18	0.65	33.0	74.8	16.1
D233	480	1.19	0.32	18	0.61	32.3	72.2	14.3
D264	400	1.19	0.32	23	0.60	29.5	65.7	11.8
D988	550	1.27	0.46	19	0.72	29.1	72.5	15.2
D915	550	1.39	0.65	21	0.78	23.6	65.5	12.2
D912	550	1.50	0.80	24	0.82	16.3	65.9	8.8

The properties studied include the substrate temperature during growth (T_SS), the Ga content, and the optical gap (E_g). The Urbach energies (E_U) as determined by the TPC spectra are included as well. Also listed are the device performance characteristics: the open-circuit voltage (V_oc), short-circuit current (J_sc), the fill factor (FF), and the efficiency (η)

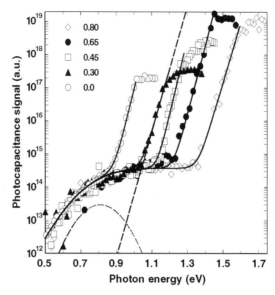

Fig. 5.3. Transient photocapacitance spectra from polycrystalline samples with Ga contents ranging from $Ga/(In + Ga) = 0$ to 0.8 have been aligned to emphasize the similarity of the deep defect response. Fits (*solid lines*) are obtained by integrating over a single Gaussian defect band, centered at 0.8 eV, plus an exponential band of tail states. Note that the Gaussian deep defect band has nearly the same appearance for all the samples and that the exponential bandtail of states have very similar slopes

most likely originates from a point defect within the CIGS alloy crystal. Such a defect level fixed relative to the valence band, independent of the Ga fraction, agrees with the behavior predicted from theory for several classes of defects in CIGS, particularly metal atom vacancies and metal antisite defects [30].

A schematic of this situation is shown in Fig. 5.4. Note that because the transition energy of this defect remains fixed with respect to the valence band, it will be located much nearer the mid gap as the Ga fraction is increased. This suggests that it would become a much more efficient recombination center in the higher Ga alloys and this provides a possible explanation for why the device performance suffers in the higher gap $CuIn_{1-x}Ga_xSe_2$ alloys [28, 31].

In addition to their quite similar defect bands near $E_V + 0.8\,eV$, all of the spectra in Fig. 5.3 exhibit bandtails with nearly the same Urbach energies, all within 4 meV of 20 meV. This is perhaps also somewhat surprising since the bandtail width in such samples is generally believed to arise from alloy disorder [13]. Indeed, we did not identify any systematic trend with Ga content at all for the CIGS samples deposited at 550°C (see Table 5.1).

In Fig. 5.5, we compare the photocapacitance for a different set of samples. Three of the samples in this case had Ga fractions near 30 at.%; however, two of these samples were deposited at lower temperatures: 480°C and 400°C in-

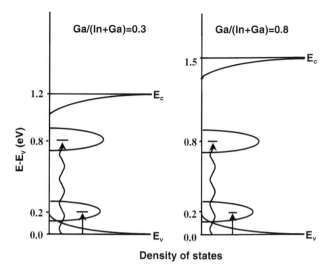

Fig. 5.4. Schematic of observed defect bands in the low and high Ga fraction CIGS alloys. The 0.8 eV optical transitions are shown by the *wavy lines*. Because the energy is relatively constant with respect to the valence band, this implies that it lies much closer to mid gap in the higher Ga fraction alloys. These samples also exhibit a thermal transition to a defect level lying within 0.3 eV of the valence band as shown by the *solid lines*

stead of the usual 550°C. The two lower-temperature samples are observed to yield spectra that are significantly different: the Urbach energies are larger (broader bandtails) and the shape of the defect band shoulder is also different. Indeed, in order to fit the defect feature for these samples two overlapping Gaussian defect bands must be incorporated. Since the photovoltaic performance of devices fabricated from such lower-temperature CIGS absorbers is generally poorer, these differences in electronic properties revealed by the photocapacitance spectra may be important to understand more fully.

One very intriguing correlation can already be identified between the photocapacitance spectra on the CIGS samples described above and the photovoltaic device performance. In Fig. 5.6, we plot the device short-circuit current density under AM1.5 illumination vs the Urbach energies for the samples with Ga fractions near 30 at.% Ga. The correlation revealed appears to be significant. While this is not understood at present, it may be related to the fact that the bandtail widths revealed by the TPC spectra indicate the density of shallow states extending from the conduction band into the gap, that these states limit the minority carrier mobilities and thus enhance the probability of recombination before the photogenerated electrons can be collected.

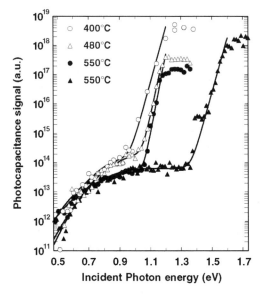

Fig. 5.5. Photocapacitance spectra for devices grown with differing substrate temperatures. The spectrum shown in *filled triangles* is from a device with Ga/(In + Ga) = 0.8; otherwise, all devices shown have Ga/(In + Ga) = 0.3. Note that the lower substrate temperature samples exhibit a band of deep defects that have a distinctly different shape than the 550°C samples

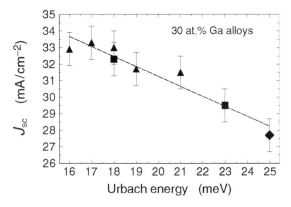

Fig. 5.6. Short-circuit current densities under AM1.5 illumination vs the photocapacitance determined Urbach energies for the samples with Ga fractions near 0.3. *Filled triangles* denote samples with $T_{SS} = 550°C$ while *filled squares* denote samples deposited below 500°C (see Table 5.1). The *single diamond* data point was obtained for an epitaxial single-crystal CIGS film deposited at the University of Illinois [25]

5.4 Determination of Minority Carrier Collection

The issue of minority carrier collection in these photovoltaic devices can be addressed more directly by experimental means. Thus far in our discussion of the photocapacitance spectra we have focused on Eq. 5.3; i.e., on optically induced transitions between the valence band and empty states in the gap. We have also assumed that for each such optical transition the remnant hole in the valence band is swept out of the depletion region, increasing its negative charge density, thus increasing the junction capacitance. This then results in a positive contribution to the photocapacitance signal as defined by Eq. 5.1. Since the time window in our measurement exceeds 100 ms, the assumption that essentially all such photogenerated valence band holes escape the depletion region is undoubtedly quite good.

In contrast, however, optically induced transitions between occupied gap states and the conduction band as given by Eq. 5.4 will lead to a *decrease* in the magnitude of negative charge in the depletion region as the photogenerated electrons are swept out. This results in a photocapacitance signal of *negative sign* [32]).[1] Actually, the situation is a bit more complex and is illustrated in Fig. 5.7. In this diagram, we have separated the types of optical transitions into four types. Type (a) is what we have focused upon thus far – the optical excitation between the valence band and unoccupied defect states with the subsequent escape of the valence band hole. Type (b) is the second type mentioned above – the removal of an electron from an occupied defect level into the conduction band with the subsequent escape of the electron. However, in p-type material such occupied levels will lie quite close to the valence band. Thus, the hole left in the defect will be quickly thermally emitted into the valence band where it then also escapes the depletion region [type (d)]. Such a case results in no change in charge state within the depletion and, hence, no photocapacitance signal. Similarly, for a transition from the valence band into an unoccupied gap state close to the conduction band, the electron will be quickly thermally emitted into the conduction band and leave the depletion region as well, with nearly zero photocapacitance signal [type (c)].

On the other hand, while transitions of types (c) and (d) result in small changes in the junction capacitance, they result in a large junction photocurrent. In general terms, if our sub-bandgap light results in n_h holes and n_e electrons leaving the depletion within the experimental time window, then the junction photocurrent signal will be proportional to $n_h + n_e$, while the photocapacitance signal will be proportional to $n_h - n_e$. Thus, it is really only by measurements of both the TPC signal as well as the transient photocurrent (TPI) signal that we can truly distinguish the quantities of holes and electrons excited out of the depletion region due to the sub-bandgap light.

We expect that transitions of types (c) and (d) will be dominant when the photon energy lies only slightly below the bandgap energy. For such transitions, each photon effectively produces one valence band hole plus one conduction band electron with no net change in the gap state occupation. However, the photocapacitance signal in this energy regime as exhibited in the TPC

---

[1]Negative photocapacitance signals are relatively rare and have only been observed in two systems to date: amorphous silicon–germanium alloys and some mixed phase amorphous/nanocrystalline Si materials.

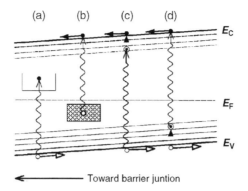

Fig. 5.7. Schematic of four different types of sub-bandgap optical transitions. Optical transitions are shown by *wavy lines*, and thermal transitions are shown by *vertical solid arrows*. *Horizontal arrows* indicate subsequent motion of the released carriers

spectra (Figs. 5.3 and 5.5) certainly does not vanish; rather, it is substantially *positive*. This indicates that photogenerated holes are more likely to escape the depletion region than photogenerated electrons. Moreover, by comparing the TPC and TPI spectra directly we can determine quantitatively the relative escape fraction, n_e/n_h.

In Fig. 5.8, we display both types of spectra for one CIGS sample. The units of these two types of spectra are different (time integrated changes in current vs capacitance); however, if we align these two spectra in the region of the defect band (optical energies below 0.9 eV), we see that the TPI spectra is then 100 times larger than the TPC spectra in the bandtail regime and that this ratio is nearly a constant.

From our discussion above, the ratio R of the two signal types will be given by

$$R = \frac{S_{TPC}}{S_{TPI}} = K\frac{n_h - n_e}{n_h + n_e} = K\frac{1 - n_e/n_h}{1 + n_e/n_h}, \qquad (5.6)$$

where K is the ratio when only one carrier type leaves the depletion region. If the photocarriers are excited uniformly over the entire depletion region and if the band bending is parabolic (valid for uniform doping and small densities of deep defects), then K can be shown to have the value $-2 \times [\text{rate window}]/(V + V_{bi})$ for a one-sided junction with built-in potential V_{bi}, and applied reverse bias V. For cases where the deep defect densities contribute substantially to the depletion charge density, the leading coefficient (of 2) will be increased, but only by a factor of order unity. The TPC and TPI signals in the defect band region of Fig. 5.8 have a ratio which, within the uncertainties of our numerical calculations, agrees quite well with the single carrier type of excitation. Thus, we believe that this portion of our sub-band gap spectra is due to transitions of type (a) in Fig. 5.7. On the other hand, we believe the nearly constant ratio R in the bandtail region directly reveals the free carrier collection ratio.

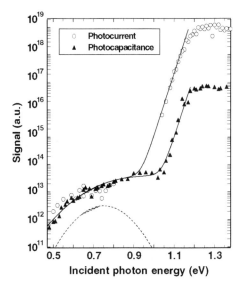

Fig. 5.8. Photocapacitance spectrum (*filled symbols*) and photocurrent spectrum (*open symbols*) for $CuIn_{0.7}Ga_{0.3}Se_2$ device D934, showing the difference between these spectra caused by minority carrier collection. Fits (*solid lines*), again result from the sum of a Gaussian band of defect states and an exponential band of tail states

Knowing R and K thus allows us to determine n_e/n_h from Eq. 5.6. For the sample in Fig. 5.8 it is nearly 0.99.

In Table 5.2, we have included the values n_e/n_h determined in this fashion for the samples where reliable TPI spectra have been obtained. Unfortunately, we have found that such TPI spectra are generally not as easy to obtain as high-quality TPC spectra. This depends on the magnitude of the dark leakage current under applied bias. Although we zero out such DC currents with our current amplifier, they still contribute noise and thus can obscure the small currents coming from the optically induced gap state transitions.

Table 5.2 also contains estimates of the magnitudes of the 0.8 eV defect transition. Reliable estimates of these magnitudes also depend on high-quality TPI spectra, particularly in the low optical energy regime. Hence, the rather large error bars. Partly for this reason perhaps, correlations between the magnitudes of this defect band and the device performance are not at all apparent. However, there does seem to be a reasonably good correlation between the device performance and the minority carrier collection fractions. In Fig. 5.9, we have plotted the photovoltaic efficiencies and fill factors vs the n_e/n_h collection fraction for the CIGS samples with roughly 30 at.% Ga fractions that were deposited at 550°C. We note, however, that the apparent correlation seen in this figure does not extend to some of the lower substrate temperature 30% Ga samples whose efficiencies were high in spite of exhibiting relatively low

Table 5.2. Summary of properties determined for selected CIGS samples using TPC and TPI spectroscopies

sample #	T_{SS} (°C)	E_g (eV)	$\frac{Ga}{Ga+In}$	E_U (meV)	N_d (a.u.)	n_e/n_h	Eff(%)
D469	550	1.0	0.0	18	20 ± 8	0.97	11.3
D362	550	1.16	0.27	16	5 ± 3	0.973	11.2
D427	550	1.18	0.29	19	5 ± 3	0.98	14.5
D934	550	1.18	0.30	21	4 ± 3	0.99	15.2
D400	550	1.20	0.32	18	5 ± 3	0.99	16.1
D233	480	1.19	0.32	18	–	0.92	14.3
D264	400	1.20	0.32	23	–	< 0.90	11.8
D988	550	1.29	0.46	19	3 ± 3	0.97	15.2
D915	550	1.42	0.65	21	4 ± 3	0.65	12.2
D912	550	1.53	0.80	24	7 ± 3	0.63	8.8

The properties studied include the Urbach energies (E_U), an estimate of the magnitudes of the 0.8 eV deep transition (N_d), and the minority carrier collection fraction (n_e/n_h)

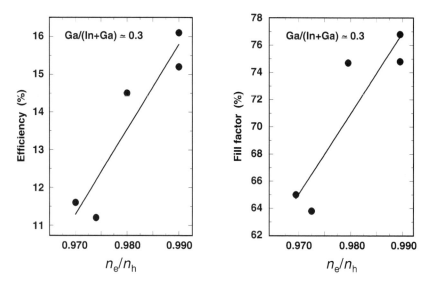

Fig. 5.9. Cell efficiencies and fill factors vs minority carrier collection ratios, n_e/n_h, for five 30% Ga fraction devices deposited at 550°C. For these samples, a fairly good correlation appears to be exhibited, primarily because of changes in fill factor. Indeed, lower fill factors are expected to result from poorer minority carrier collection [34]

electron collection efficiencies. Clearly then, additional factors undoubtedly also play a significant role in determining the overall device performance.

In addition, we have found that the n_e/n_h collection fraction becomes significantly lower for sample devices whose Ga content exceeds 40 at.%. This is almost precisely where the device efficiencies begin to drop below the predicted values [28,31]. Thus, we believe the minority/majority carrier collection fractions determined in this fashion are likely strongly linked to the performance of CIGS photovoltaic devices. On the other hand, although we are able to measure these collection fractions, we do not yet know which aspects of the electronic structure of CIGS most directly affect their values.

5.5 Photocapacitance Spectra in the Cu(InAl)Se$_2$ Alloys

Recently, we carried out the first evaluation of the electronic properties of a series of CuIn$_{1-x}$Al$_x$Se$_2$ (CIAS) alloy sample devices [35]. These samples were fabricated at IEC using a process very similar to those developed to optimize CIGS devices as has been described above [36, 37]. X-ray diffraction studies have verified that these CIAS films are single phase. The devices for this study contained CIAS absorber layers with Al fractions ranging from $x = 0.13$ to 0.48 as determined by energy dispersive X-ray spectroscopy. The device performance parameters were obtained at IEC for all devices, and ranged from 13% to 7%, with device efficiency decreasing as the Al content of the absorber layer increased. The details of the film parameters and device properties are given in Table 5.3.

Replacing In with Al in CIS enables the bandgap to be increased using much smaller alloying fractions; typically, a given Al fraction increases the bandgap as much as twice the Ga fraction. Moreover, for Al fractions below 20 at.% quite good results were obtained for the photovoltaic device performance. Figure 5.10 compares the quantum efficiencies (QE) of two devices – one using a 26 at.% Ga CIGS absorber and the other using a 13 at.% Al CIAS absorber [38]. Both had essentially identical junction contacts. The QE curves are essentially identical.

Sub-bandgap optical spectra of the film properties within the devices were obtained from TPC spectroscopy. A comparison of a similar 13 at.% Al CIAS sample device with a 27 at.% Ga CIGS sample device is displayed in Fig. 5.11.

Table 5.3. CIAS devices studied, in order from lowest to highest Al content, together with their properties

sample #	E_g (eV)	$\dfrac{Al}{Al+In}$	E_U (meV)	V_{oc} (V)	J_{sc} (mA/cm^2)	FF (%)	Eff. (%)
D242	1.15	0.13	17	0.59	34.2	65	13.1
D287	1.36	0.29	27	0.71	25.3	63	11.3
D289	1.45	0.35	42	0.72	21	64	9.6
D303	1.67	0.48	35	0.72	16.9	63	6.3

The properties studied include the Al content, optical gap (E_g), and the device performance parameters. Urbach energies (E_U) as determined by the TPC spectra are also included

84   J.D. Cohen et al.

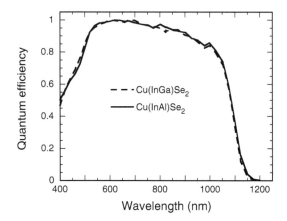

Fig. 5.10. Quantum efficiency curves for a CIGS device with 26 at.% Ga and a CIAS device with 13 at.% Al. Total efficiencies under AM1.5 illumination were 16.5% and 16.9%, respectively

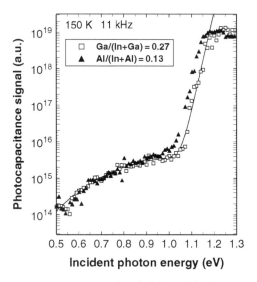

Fig. 5.11. Photocapacitance spectra for CIGS and CIAS devices similar to those in Fig. 5.10 (devices D362 and D242). The CIGS sample had a slightly larger Ga fraction

We deduce that two samples have identical Urbach energies, defect bands and, from the magnitudes of signals in the bandtail, nearly identical minority carrier collection fractions. The bandgap in this case is slightly larger for the CIGS sample; however, this is because the Ga fraction was also a bit higher than for the CIGS sample used to compare the device performance in Fig. 5.10. Thus, the two photocapacitance spectra are nearly the same for two samples whose quantum efficiencies and device performance are nearly

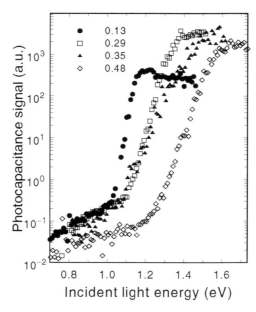

Fig. 5.12. Photocapacitance spectra for four CIAS samples with different Al fractions as indicated. Note that there is a marked increase in the Urbach energies as Al content increases beyond 20 at.%, and also that the defect band becomes broader and more featureless. The smaller dynamic range of the bandtail for the 0.13 Al fraction film indicates a higher minority carrier collection efficiency for this sample

identical. This gives considerable credence to the relevance of our TPC characterization method toward understanding how the corresponding devices are likely to behave.

The TPC spectra for all four of the CIAS devices studied to date are shown in Fig. 5.12. Here, we see that the electronic properties of these alloys change quite dramatically once the Al fraction exceeds 20 at.%. Specifically, the Urbach energies (bandtail widths) become much broader, probably indicating a much larger degree of alloy disorder. The characteristic shape of the defect band "shoulder" appears significantly flatter for the higher Al fraction alloys. We also note that the bandtail extends over fewer orders of magnitude in the devices with the lowest Al fraction and believe this is probably a result of better minority carrier collection in the lower aluminum fraction devices.

The electronic properties of these samples, as deduced from these photocapacitance spectra are listed in Table 5.3 along with their device performance parameters. The broader Urbach energies certainly account, at least in part, for the lower than expected device open-circuit voltages for the higher-bandgap CIAS alloys. This reduction in V_{oc} with increased bandtailing has been discussed in detail in the case of amorphous silicon device performance [39, 40].

5.6 Conclusions

We have presented a range of experimental results obtained using a method that is relatively new to the study of CIGS alloys – the TPC method. The sub-bandgap optical spectra obtained from this method have allowed us to determine Urbach energies, minority carrier collection efficiencies and have identified a new defect band associated with a 0.8 eV optical transition from the valence band. In addition, in sample devices deposited below 500°C this technique has indicated a somewhat more complex set of deep defect transitions. The fact that all of these measurements have been carried out directly on working photovoltaic devices means that we could immediately attempt to correlate the differences observed in the electronic structure with differences in the photovoltaic device performance. Such comparisons have suggested some intriguing possible causes for the poorer performance exhibited in many cases. However, we do not believe at this point that any of the correlations noted in these sections actually prove a direct causal link. Rather, they point to plausible explanations that could never have even been considered previously because such aspects of the electronic structure had never been measured. It is worthwhile to summarize these here, therefore, simply in the form of proposals or hypotheses.

First, we were able to detect a distinct band of deep defects in the CIGS alloys as revealed by an optical transition between this set of unoccupied defect levels and the valence band. Because the associated optical energy threshold does not change systematically with Ga content, this implies that this defect band approaches mid gap as the Ga fraction is increased in these alloys. Thus, it is likely to become a much more efficient recombination center. We propose that this offers at least one possible explanation for why the efficiencies of CIGS devices drop off considerably for Ga fractions above 40 at.%.

Second, in CIGS alloy samples deposited below 500°C, our TPC spectra reveal a more complex defect band structure that appears to consist of more than the single broad defect band observed in the higher deposition temperature sample devices. An interesting question, therefore, is whether this change in the structure of the deep defects is linked to the poorer device performance observed in the lower-temperature samples.

Third, our spectra provide us with a direct measurement of the bandtail parameter, or Urbach energy. Larger Urbach energies indicate broader bandtails, and these are believed to primarily indicate the degree of alloy disorder in the CIGS materials. Somewhat surprisingly, for the samples deposited at 550°C, the Urbach energies are largely *uncorrelated* with the Ga fractions as these vary between 0 and 80 at.%. However, they were found to be significantly larger for a 400°C deposited (smaller grain size) CIGS sample. For the sample devices with Ga fractions close to 30 at.%, we identified a fairly significant inverse correlation between the device short-circuit current density and the Urbach energy. This is still not well understood, although it might be linked to an increased degree of minority carrier trapping in a broader conduction bandtail and this may ultimately increase the likelihood of recombination.

Fourth, in our study of four CIAS samples, for which Al was used in place of Ga to increase the bandgap, we found that the electronic properties were quite similar in all respects to the CIGS samples at lower Al levels. This was consistent with the nearly identical device performance exhibited for CIAS and CIGS devices of similar bandgaps below about 1.2 eV. However, we found that the Urbach energy increased dramatically for Al fractions above 20 at.%. This contrasts with the higher-gap materials produced by Ga alloying for which the Urbach energies exhibited only relatively small changes. Our results for the CIAS alloys thus suggest additional deterioration in electronic properties as the bandgap is increased. We believe this would account, at least in part, for the loss in open-circuit voltage that such CIAS-based devices are observed to exhibit. However, it is not yet clear whether such poorer electronic properties are intrinsic to these CIAS alloys, or may ultimately be overcome.

Finally, we demonstrated that by comparing TPC spectra with transient junction photocurrent measurements in the same samples, we could estimate the minority carrier (electron) collection fraction following generation by near-bandgap illumination. Moreover, for a series of CIGS alloys with 30 at.% Ga fractions, we found that higher electron collection fractions measured in this manner correlated fairly well with the device efficiencies, primarily due to higher fill factors. This analysis also indicates that the higher-bandgap alloys generally exhibit lower electron collection, consistent with their lower than expected device efficiencies. At present, however, we have not identified a specific aspect of the electronic structure of these alloys that is responsible for such lowered collection fractions.

Therefore, our TPC studies have raised as many questions as they have answered. However, we believe that without the types of information these measurements have provided, attempts to correlate cell performance with electronic properties of their CIGS absorbers could never have progressed very far. Moreover, we have been greatly encouraged by the observed similarity of the photocapacitance spectra for sample devices that have similar performance parameters. Serious attempts to model these devices incorporating the features deduced by our photocapacitance measurements have already begun [41]. Taken together with the information provided by admittance-based measurements, we believe that efforts to understand the detailed relation between device performance and electronic properties of these chalcopyritye materials are quite likely to succeed in the near future.

References

1. Ramanathan, K., Contreras, M.A., Perkins, C.L., Asher, S., Hasoon, F.S., Keane, J., Young, D., Romero, M., Metzger, W., Noufi, R., Ward, J., Duda, A.: Properties of 19.2% efficiency ZnO/CdS/CuInGaSe2 thin-film solar cells. Prog. Photovolt: Res. Appl. 11, 225–230 (2003)

2. Walter, T., Herberholz, R., Müller, C., Schock, H.W.: Determination of defect distributions from admittance measurements and application to Cu(In,Ga)Se$_2$ based heterojunctions. J. Appl. Phys. 80, 4411–4420 (1996)

3. Herberholz, R., Igalson, M., Schock, H.W.: Distinction between bulk and interface states in CuInSe$_2$/CdS/ZnO by space charge spectroscopy. J. Appl. Phys. 83, 318–325 (1998)

4. Rau, U., Braunger, D., Herberholz, R., Schock, H.W., Guillemoles, J.-F., Kronik, L., Cahen, D.: Oxygenation and air-annealing effects on the electronic properties of Cu(In,Ga)Se$_2$ films and devices. J. Appl. Phys. 86, 497–505 (1999)

5. Turcu, M., Ktschau, I.M., Rau, U.: Composition dependence of defect energies and band alignments in the Cu(In$_{1-x}$Ga$_x$)(Se$_{1-y}$S$_y$)$_2$ alloy system. J. Appl. Phys. 91, 1391–1399 (2002)

6. Igalson, M., Schock, H.W.: The metastable changes of the trap spectra of CuInSe$_2$-based photovoltaic devices. J. Appl. Phys. 80, 5765–5769 (1996)

7. AbuShama, J., Johnston, S., Ahrenkiel, R., Noufi, R.: Deep level transient spectroscopy and capacitance-voltage measurements of Cu(In,Ga)Se$_2$. Proceedings of the 29th IEEE Photovoltaic Specialists Conference, pp. 740–743, (2002)

8. Michelson, C.E., Gelatos, A.V., Cohen, J.D.: Drive-level capacitance profiling: its application to determining gap state densities in hydrogenated amorphous silicon films. Appl. Phys. Lett. 47, 412–414 (1985)

9. Heath, J.T., Cohen, J.D., Shafarman, W.N.: Distinguishing metastable changes in bulk CIGS defect densities from interface effects. Thin Solid Films 431–432, 426–430 (2003)

10. Heath, J.T., Cohen, J.D., Shafarman, W.N.: Bulk and metastable defects in CuIn$_{1-x}$Ga$_x$Se$_2$ thin films using drive-level capacitance profiling. J. Appl. Phys. 95, 1000–1010 (2004)

11. Rau, U., Weinert, K., Nguyen, Q., Mamor, M., Hanna, G., Jasenek, A., Schock, H.W.: Device analysis of Cu(In,Ga)Se$_2$ heterojunction solar cells - some open questions. Mater. Res. Soc. Symp. Proc. 668, H9.1.1-H9.1.12 (2001)

12. Cody, G.D., Tiedje, T., Abeles, B., Brooks, B., Goldstein, Y.: Disorder and the optical-absorption edge of hydrogenated amorphous silicon. Phys. Rev. Lett. 47, 1480–1483 (1981)

13. Wasim, S.M., Rincon, C., Marin, G., Bocaranda, P., Hernandez, E., Bonalde, I., Medina, E.: Effect of structural disorder on the Urbach energy in Cu ternaries. Phys. Rev. B 64, 195101–195108 (2001)

14. Orenstein, J., Kastner, M.A.: Time-resolved optical absorption and mobility of localized charge carriers in a-As$_2$Se$_3$. Phys. Rev. Lett. 43, 161–165 (1979)

15. Tiedje, T., Rose, A.: A physical interpretation of dispersive transport in disordered semiconductors. Solid State Commun. 37, 49–52 (1981)

16. Medvedkin, G.A., Rud, Y.V., Tairov, M.A.: Optical band edge absorption in CuInSe$_2$. Phys. status solidi B 144, 809–816 (1987)

17. Shioda, T., Chichibu, S., Irie, T., Nakanishi, H.: Influence of nonstoichiometry on the Urbach's tails of absorption spectra for CuInSe$_2$ single crystals. J. Appl. Phys. 80, 1106–1111 (1996)

18. Jackson, W.B., Amer, N.M.: Direct measurement of gap-state absorption in hydrogenated amorphous silicon by photothermal deflection spectroscopy. Phys. Rev. B 25, 5559–5562 (1982)

19. Vanecek, M., Kocka, J., Stuchlik, J., Kozisek, Z., Stika, O., Triska, A.: Density of the gap states in undoped and doped glow discharge a-Si:H. Sol. Energy Mater. 8, 411–423 (1983)

20. Meeder, A., Fuertes Marron, D., Rumberg, A., Lux-Steiner, M.Ch., Chu, V., Conde, J.P.: Direct measurement of Urbach tail and gap state absorption in CuGaSe$_2$ thin films by photothermal deflection spectroscopy and the constant photocurrent method. J. Appl. Phys. 92, 3016–3020 (2002)

21. Guillemoles, J.-F., Kronik, L., Cahen, D., Rau, U., Jasenek, A., Schock, H.W.: Stability issues of Cu(In,Ga)Se$_2$-based solar cells. J. Phys. Chem. B 104, 4849–4862 (2000)

22. Gelatos, A.V., Cohen, J.D., Harbison, J.P.: Assessment of lattice relaxation effects in transitions from mobility gap states in hydrogenated amorphous silicon using transient photocapacitance techniques. Appl. Phys. Lett. 49, 722–724 (1986)

23. Gelatos, A.V., Mahavadi, K.K., Cohen, J.D., Harbison, J.P.: Transient photocapacitance and photocurrent studies of undoped hydrogenated amorphous silicon. Appl. Phys. Lett. 53, 403–405 (1988)

24. Heath, J.T., Cohen, J.D., Shafarman, W.N., Johnson, D.C.: Characterization of Deep Defects in CuIn$_{1-x}$Ga$_x$Se$_2$ (CIGS) Working Photovoltaic Devices. Electrochem. Soc. Conf. Proc. 10, 324–332 (2001)

25. Heath, J.T., Cohen, J.D., Shafarman, W.N., Liao, D.X., Rockett, A.A.: Effect of Ga content on defect states in CuIn$_{1-x}$Ga$_x$Se$_2$ photovoltaic devices. Appl. Phys. Lett. 80, 4540–4542 (2002)

26. Lang, D.V.: Deep-level transient spectroscopy: a new method to characterize traps in semiconductors. J. Appl. Phys. 45, 3023–3032 (1974)

27. Lang, D.V.: In: Bräunlich, P. (ed.) Thermally Stimulated Relaxation in Solids. Topics in Applied Physics, vol. 37, pp. 93–134. Springer, Berlin Heidelberg New York (1979)

28. Shafarman, W.N., Klenk, R., McCandless, B.E.: Device and material characterization of Cu(InGa)Se$_2$ solar cells with increasing band gap. J. Appl Phys. 79, 7324–7328 (1996)

29. Shafarman, W.N., Zhu, J.: Effect of grain size, morphology and deposition temperature on Cu(InGa)Se$_2$ solar cells. Mater. Res Soc. Symp. Proc. 668, H2.3.1–H2.3.6 (2001)

30. Wei, S.H., Zhang, S.B., Zunger, A.: Effects of Ga addition to CuInSe$_2$ on its electronic, structural, and defect properties. Appl. Phys. Lett. 72, 3199–3201 (1998)

31. Rau, U., Schock, H.W.: Electronic properties of Cu(In,Ga)Se$_2$ heterojunction solar cells-recent achievements, current understanding, and future challenges. Appl. Phys. A 69, 131–147 (2002)

32. Chen, C.-C., Zhong, F., Cohen, J.D., Yang, J.C., Guha, S.: Evidence for charged defects in intrinsic glow-discharge hydrogenated amorphous-silicon–germanium alloys. Phys. Rev. B 57, R4210–R4213 (1998)

33. Kwon, D., Chen, C.-C., Cohen, J.D., Jin, H.-C., Hollar, E., Robertson, I., Abelson, J.R.: Electronic transitions associated with small crystalline silicon inclusions within an amorphous silicon host. Phys. Rev. B 60, 4442–4445 (1999)

34. Eron, M., Rothwarf, A.: Effects of a voltage-dependent light-generated current on solar cell measurements: CuInSe$_2$/Cd(Zn)S. Appl. Phys. Lett. 44, 131–133 (1984)

35. Heath, J.T., Cohen, J.D., Shafarman, W.N.: Defects in copper indium aluminum diselenide films and their impact on photovoltaic device performance. Mater. Res. Soc. Symp. Proc. 763, 441–446 (2003)

36. Shafarman, W.N., Marsillac, S., Paulson, P.D., Haimbodi, M.W., Minemoto, T., Birkmire, R.W.: Material and Device Characterization of Thin Film Cu(In,Al)Se$_2$ Solar Cells. In: Proceedings of the 29th IEEE Photovoltaic Specialists Conference2002, pp. 519–522. IEEE, New Orleans (2002)
37. Shafarman, W.N., Klenk, R., McCandless, B.E.: Device and material characterization of Cu(InGa)Se$_2$ solar cells with increasing band gap. J. Appl. Phys. 79, 7324–7328 (1996)
38. Marsillac, S., Paulson, P.D., Haimbodi, M.W., Birkmire, R.W., Shafarman, W.N.: High-efficiency solar cells based on Cu(InAl)Se$_2$ thin films. Appl. Phys. Lett. 81, 1350–1352 (2002)
39. Tiedje, T.: Band tail recombination limit to the output voltage of amorphous silicon solar cells. Appl. Phys. Lett. 40, 627–629 (1982)
40. Jiang, L., Lyou, H.H., Rane, S., Schiff, E.A., Wang, Q., Yuan, Q.: Open-circuit voltage physics of amorphous silicon based solar cells. Mater. Res. Soc. Symp. Proc. 609, A18.3.1–A18.3.11 (2000)
41. Rockett, A.A.: Performance-Limitations in Cu(In,Ga)Se$_2$-Based Heterojunction Solar Cells. In: Proceedings of the 29th IEEE Photovoltaic Specialists Conference—2002, pp. 587–591. IEEE, New Orleans (2002)

6

Recombination Mechanisms in Cu(In,Ga)(Se,S)$_2$ Solar Cells

M. Turcu and U. Rau

An improved understanding on the recombination mechanisms in Cu(In,Ga)(Se,S)$_2$ solar cells is required to enhance the performance of these devices. This chapter first reviews the experimental work that has been undergone in the past to investigate the recombination mechanisms in Cu(In,Ga)(Se,S)$_2$ of different compositions. Then, we discuss analytical and numerical modeling of ZnO/CdS/Cu(In,Ga)(Se,S)$_2$ heterojunction solar cells and discuss the differences between the wide-gap devices and their low-gap counterparts. It is shown that a major role in the suppression of interface recombination is played by the Cu-poor surface defect layer that forms in Cu-poor chalcopyrites and leads to surface band gap widening. However, the beneficial effect of the surface defect layer is reduced when increasing the band gap without decreasing the valence band edge. The improvement due to the surface defect layer can be explained by either n-doping within the layer or by a Fermi-level pinning effect. The n-doping situation is somewhat beneficial in that it allows for higher excess defect concentrations. Based on realistic material properties a maximum efficiency of wide-gap Cu(In,Ga)Se$_2$ (CIGS) devices of 15.5% is expected due to limitations by interface recombination and the unfavorable band alignment.

6.1 Introduction

The Cu(In,Ga)(Se,S)$_2$ chalcopyrite alloy system provides absorber materials for the most efficient thin-film photovoltaic technology to date [1]. Thin-film ZnO/CdS/chalcopyrite solar cells are especially outstanding due to their low cost [2], radiation hardness [3], long-term stability [4], and record power conversion efficiencies of over 19% [5]. These record cells are made from Cu(In$_{1-x}$Ga$_x$)Se$_2$ alloys with a Ga content $x \approx 0.1 - 0.2$ and, accordingly, a band gap energy $E_g \approx 1.1 - 1.2\,\mathrm{eV}$. However, wide-gap Cu(In$_{1-x}$Ga$_x$)(Se$_{1-y}$S$_y$)$_2$ alloys with considerably enhanced Ga and/or S content would have some advantages with respect to the standard low-gap material because of their larger E_g, a higher open-circuit voltage V_{OC}, and a lower short-circuit current density j_{sc}, thus minimizing ohmic losses in thin-film modules [6–8]. However, the delivered open-circuit voltage increases at a lower rate than the band gap energy of the absorber for wide-gap devices resulting in a lower power conversion efficiency of wide-gap $Cu(In,Ga)(Se,S)_2$ solar cells when compared to their low-gap counterparts. The physical reasons for this reduced efficiency are not completely understood up to now. Therefore, this chapter provides a review on the experimental recombination analysis and the modeling of $Cu(In,Ga)(Se,S)_2$ solar cells with a special focus on wide-gap chalcopyrite absorber materials.

6.2 Review of Experimental Recombination Analysis

6.2.1 CuInSe$_2$ Cells

Early CuInSe$_2$ solar cells consisted of a thick CdS layer as window material [1]. Upon the development of device technology, the initially thick CdS window was replaced by a combination of a thin CdS buffer layer and a highly conductive ZnO window layer [1]. Naturally, the modification of the device structure induced changes in the recombination mechanisms contributing to the open-circuit voltage loss of the specific devices.

An early analysis of the loss mechanisms determining the open-circuit voltage and short-circuit current losses voltage of polycrystalline solar cells (with emphasis to CuInSe$_2$) was made by Sites in 1988 [9]. This early analysis, based on devices with efficiencies of around 10%, already points to the major limitations that are still operative on the much higher efficiency level of the present CIGS devices. The relatively low doping level of the order of few $10^{15}\,\mathrm{cm}^{-3}$ in CuInSe$_2$ due to the high level of compensation is responsible for a low built-in potential and a relatively wide space charge region (SCR). Thus, recombination via defects in the SCR results as the dominating loss mechanism for the open-circuit voltage. The short-circuit current was found to be limited by optical losses in the thick CdS window layer by reflection and by collection losses in the absorber. The fill factor was limited by the diode ideality factor close to 2 and by series resistance effects. From his analysis, Sites predicted an optimization potential of CuInSe$_2$ up to 15% [9], a level that was in fact realized less than a decade later.

Roy et al. [10] have performed temperature-dependent current–voltage analysis under illumination on early CuInSe$_2$/(CdZn)S devices with different Zn concentrations (0%, 12%, and 26%) in the (CdZn)S layer. The analysis of the main heterojunction revealed an activation energy equal to the bandgap of CuInSe$_2$ and independent of the bandgap of (CdZn)S buffers. The experimental findings were taken as indication that forward conduction is dominated by recombination in the SCR of CuInSe$_2$ in those devices [10]. Another analysis of current–voltage characteristics of early CuInSe$_2$/(CdZn)S devices (with 10% Zn in (CdZn)S), performed by Shafarman and Phillips [11], indicated that the specific cells are also controlled by recombination within the CuInSe$_2$ absorber.

When investigating early all-sputtered CdS/CuInSe$_2$ solar cells, Santamaria et al. [12] have found a current transport mechanism dominated either by tunneling or by interface recombination. In such solar cells, the major contribution came from interface recombination mechanism for devices based on stoichiometric or indium-rich CuInSe$_2$, and from tunneling conduction mechanism for copper-rich devices [12].

Phillips et al. [13] performed a representative analysis of temperature dependence of the diode current in polycrystalline heterojunction solar cells (including CuInSe$_2$ and Cu(In,Ga)Se$_2$). The authors concluded that the mechanism controlling the recombination current in the CuInSe$_2$/CdS solar cells under investigation (and in respective CIGS-based devices) is the recombination in the SCR of the absorber [13].

When performing analysis of early ZnO/CdS/CuInSe$_2$ solar cell, Mitchell and Liu [14] concluded that recombination controls device performance above 200 K, while tunneling and series resistance dominate low-temperature device behavior. Yoo et al. [15] have found evidence for recombination in the depletion region with or without thermally assisted tunneling as dominant transport mechanisms in the dark for the early ZnO/CdS/CuInSe$_2$ solar cell.

Walter et al. [16] have considered recombination in the SCR through single defects and have extended the recombination model to the case of an exponential increase of the density of defects towards the valence band edge. The model of Walter et al. has been later generalized by Rau [17] to the case of tunneling-enhanced recombination via trap states in the SCR. This model consistently fits the experimental current/voltage data of ZnO/CdS/Cu(In,Ga)Se$_2$ solar cells in a temperature range between $T = 350$ and 100 K. The tunneling contribution to the recombination is negligible at room temperature but becomes significant at $T < 250$ K. In the same paper it was found that increasing the Ga content in the CIGS alloy leads to enhanced tunneling and that in CuGaSe$_2$ this contribution is dominant already at room temperature.

In summarizing more than 15 years of investigation of the electronic transport in CuInSe$_2$ and Cu(In,Ga)Se$_2$ solar cells, we find that qualitatively all authors have come to the same conclusion: bulk recombination and, hence, the bulk properties of the absorber material determine the open-circuit voltage. The dramatic efficiency improvements during the last decade, therefore, primarily result from optimizing the electronic quality of the absorber. Key parameters for this recent achievement as well as for further possible improvements are the concentration of recombination centers and the free carrier density. The free carrier density determines the width of the SCR as well as the built-in potential. Increasing this value to a level of 10^{17} cm^{-3} would decrease SCR recombination by lowering the width of the SCR. The same improvement would also decrease recombination in the neutral zone of the absorber by the consequent increase of the built-in potential. An upper limit for the beneficial effect of a high doping density is imposed by tunneling that occurs at higher charge densities in the SCR. Unfortunately, even in the best CIGS devices today, the high degree of compensation does not allow adjustment of the free carrier density deliberately. In addition, the number of deep, recombination active defects should be as low as possible. Thus, a key issue for a systematic improvement of CIGS-based heterojunction solar cells is a better understanding of the defect physics of this material, in terms of the origin of the defects and their consequences on the device performance.

6.2.2 Wide-Gap Chalcopyrite Cells

The bandgap energy of the absorber appears to be an important element when considering the relative contributions of recombination in the bulk of the absorber and at the CdS/absorber interface in ZnO/CdS/chalcopyrite devices [6]. The increase of the bandgap energy (E_g) following an increase of the Ga content x in the $Cu(In_{1-x}Ga_x)Se_2$ absorber is accommodated almost exclusively by an increase of the energy E_C of the conduction band minimum. As a consequence, the barrier for interface recombination at the $CdS/Cu(In_{1-x}Ga_x)Se_2$ remains essentially unchanged with respect to the value for low-Ga absorbers. Hence, the relative contribution of interface recombination should increase with increasing Ga content. Therefore, it was argued [6] that interface recombination is responsible for the moderate performance of wide-gap chalcopyrite solar cells.

Malmström et al. [18] have performed temperature-dependent current–voltage analysis on a set of actual $CuIn_{1-x}Ga_xSe_2$-based solar cells with different Ga contents in the absorber. The authors have used the one-diode model to fit the measured characteristics of the devices and the temperature dependence of the model parameters was used to draw conclusions on the forward current transport mechanism [18]. While the devices with $x \leq 0.5$ were found to be dominated by tunneling-enhanced recombination in the SCR, tunneling-enhanced recombination at the CdS/absorber interface was found dominant in the device with $x = 1$ [18].

Another investigation of the temperature dependence of current–voltage characteristics of different $ZnO/CdS/CuIn_{1-x}Ga_xSe_2$ solar cells is included in [19]. The data are obtained from analysis of the illuminated current–voltage characteristics by plotting short-circuit current vs open-circuit voltage for different illumination intensities [19]. Recombination in solar cells based on absorbers with Ga contents $x = 0$, 0.28, and 1 was found to take place in the volume of the absorber [19]. It was also observed that with increasing Ga content into the absorber, the contribution of tunneling increased in the investigated cells [19].

Nadenau et al. [20] have investigated electronic transport mechanisms of $CuGaSe_2$-based thin-film solar cells using a matrix of samples with different substrates (Na-free and Na-containing), different types of stoichiometry deviations (Cu-rich and Ga-rich), and different types of CdS recipes (deposition temperature 60°C or 80°C). Tunneling-enhanced *interface* recombination was proposed as the most probable dominant transport mechanism in the investigated Cu-rich CuGaSe₂ devices [20]. CuGaSe₂ cells based on Ga-rich absorbers were best described by the model for tunneling-enhanced recombination in the SCR [20].

Saad and Kassis [21] have analyzed the current–voltage characteristics of ZnO/CdS/CuGaSe₂ single-crystal solar cells depending on illumination intensity. The recombination of carriers at the interface between CdS and CuGaSe₂ was found to dominate the current transport for $V < 0.8$ V, while recombination in the depletion region was found to play a major role for $V > 0.8$ V [21].

The present authors have investigated the recombination mechanism in Cu(In,Ga)(Se,S)₂-based thin-film solar cells and the interrelationship with absorber composition [22,23]. This work used the illuminated current–voltage characteristics at different temperatures and analyzed the dependence of the open-circuit voltage from temperature to find the activation energies E_a of the recombination process. The results as shown in Fig. 6.1 indicate that devices based on absorbers with a Cu-rich composition prior to heterojunction formation are dominated by recombination at the interface with an activation energy that does not depend on the increase of E_g when achieved by increasing the S content in the alloy. In contrast, bulk recombination is dominant in devices based on Cu-poor absorbers and the activation energies roughly follow the band gap energy, regardless of whether the increase of E_g is caused by increasing the Ga/(Ga + In) or the S/(S + Se) ratio [22]. Thus, the borderline between bulk and interface recombination should be discussed in terms of Cu-poor and Cu-rich absorber compositions. Thus, the absence of interface recombination, despite the presence of nonlattice matched CdS/Cu(In,Ga)Se₂

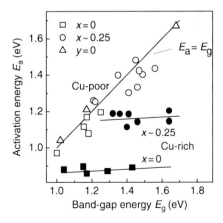

Fig. 6.1. Activation energy E_a of the dominant recombination process as a function of the bandgap energy E_g of Cu-poor (*open symbols*) and Cu-rich (*filled symbols*) Cu(In$_{1-x}$Ga$_x$)(Se$_{1-y}$S$_y$)₂ absorbers with various y and with $x = 0$ (*squares*) and $x \approx 0.25$ (*circles*) according to [22, 23]

heterointerface, is a key issue for the excellent photovoltaic performance of Cu-poor CIGS films. As we will discuss in more detail later, the enlargement of the band gap energy at the surface for Cu-poor absorber films eclipses interface recombination in these Cu-poor devices.

Reiß and coworkers [24] have analyzed the current transport in $CuInS_2$-based solar cells, where the absorbers were prepared by different sequential processes and the secondary Cu_xS phase was removed by cyanide etching. Tunneling-enhanced recombination via traps in the SCR was considered as the main recombination mechanism in the dark [23]. Under illumination, the dominant recombination in $CuInS_2$/CdS/ZnO cells was assumed to take place at the absorber/buffer interface [23]. Hengel et al. [25] have investigated the recombination mechanism in $CuInS_2$:Ga/CdS/ZnO solar cells. A change from tunneling into interface states in the dark to a thermally activated process under illumination was observed [24]. In the latter case the recombination occurred over a reduced barrier and was assumed to take place at the interface [24]. All these experimental findings are compatible with numerical simulation by Klenk [26]. There the minimization of interface recombination and criteria for efficient chalcopyrite heterojunction solar cells are discussed. According to this analysis [25], efficient heterojunction solar cells can be achieved even in the presence of a high interface recombination velocity (and in the absence of a surface bandgap widening) if the interface is inverted, i.e., if the Fermi level at the interface is above the mid gap and the barrier reduction is avoided. An inverted interface is discussed in relation to a suitable doping of the window and absorber, to an appropriate electrical charge at the interface, and to the position of the charge neutrality level when the Fermi level is pinned [25].

6.3 Role of Surface Band Gap Widening

The enlargement of the bandgap energy of the absorber towards the film surface is an important particularity of the completed device. In 1993, Schmid et al. [27] showed that free surfaces of Cu-poor grown CIGS polycrystalline thin films exhibit two prominent features:

1. The Fermi level lies above the valence band energy E_V by about 1.1 eV for $CuInSe_2$ films. This energy is larger than the bandgap energy E_g of the bulk absorber material. This finding was taken as an indication for a widening of the absorber bandgap towards the surface of the film. In 2002, a direct measurement of the surface bandgap of polycrystalline $CuInSe_2$ films by Morkel et al. [28] proved that at the surface the bandgap is about 1.4 eV, i.e., more than 0.3 eV larger than the bandgap in the bulk.
2. The surface composition of Cu-poor CIGS films corresponds to a (Ga + In)/(Cu + Ga + In) ratio of approximately 0.75 for a range of bulk compositions of $0.5 < (Ga + In)/(Cu + Ga + In) < 0.75$ [29].

Both observations (1) and (2) have led to the assumption that a phase segregation of Cu(In,Ga)$_3$Se$_5$, the so-called surface defect layer (SDL), occurs at the surface of the films [26]. The question whether the type inversion of the CIGS surface of as-grown films as well as in completed heterojunction devices results from an n-type doping of this surface layer [30–32] (doping model) or from shallow surface donors that pin the Fermi level close to the conduction band [33, 34] (pinning model) is still not finally cleared.

The next section describes a quantitative model [23] for the band diagram of ZnO/CdS/Cu(In,Ga)(Se,S)$_2$ heterostructures and for carrier recombination at the CdS/absorber interface. The model allows derivation of an analytical expression for the open-circuit voltage of the device when the recombination at the CdS/absorber interface is dominant. The analytical calculations are in relatively good agreement with numerical simulations. In Sect. 6.5, numerical simulations are further used to investigate the role of the Cu-poor surface defect layer on CIGS thin films for the photovoltaic performance of ZnO/CdS/Cu(In,Ga)Se$_2$ heterojunction solar cells [35]. We model the surface layer either as a material that is n-type doped or as a material that is type-inverted due to Fermi-level pinning by donor-like defects at the interface with CdS. We further assume a bandgap widening of this layer with respect to the CIGS bulk. This feature turns out to represent the key quality of the Cu-poor CIGS surface as it prevents recombination at the absorber/CdS buffer interface.

6.4 Analytical Model for the Heterointerface

The contents of this section have been previously presented in [23]. The starting point is an electrostatic model that describes the pinning behavior of the Fermi level at the active interface of ZnO/CdS/chalcopyrite heterojunctions under equilibrium [33]. Following the same guidelines, we extend the analytical formulation to the case of nonequilibrium conditions under applied bias and consider in addition the surface defect layer (SDL), which naturally segregates in the uppermost surface region of Cu-poor chalcopyrite thin films [26]. Figure 6.2 shows the energy band alignment of the ZnO/CdS/wide-gap Cu(In,Ga)(Se,S)$_2$ heterojunction under applied voltage V and defines the relevant quantities.

We assume an absorber material with a bulk bandgap energy E_g, a uniform p-type doping density N_a and a space charge width w_a, a completely depleted CdS buffer layer of thickness d_b and n-type doping density N_b, and a highly doped ZnO window material with negligible space charge width. The surface defect layer occurring only in the case of Cu-poor devices enlarges the absorber bandgap towards the active interface [26, 27], and therefore we consider an internal valence band offset ΔE_V^i between the SDL and the chalcopyrite (no corresponding conduction band offset is assumed).

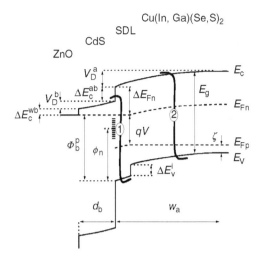

Fig. 6.2. Schematic energy band diagram of the ZnO/CdS/wide-gap Cu(In,Ga)(Se,S)$_2$ heterostructure under applied bias V and relevant recombination paths at the CdS/absorber interface (*path 1*) and in the SCR of the absorber (*path 2*). The quantities E_C, E_V, E_{Fn} and E_{Fp} denote the conduction band edge, valence band edge, electron energy and hole Fermi energy, respectively, E_g is the bandgap energy of the absorber, and ζ is the energy difference between the hole Fermi level and the valence band edge of the neutral bulk absorber. The energy distances ΔE_{Fn} and Φ_b^p determine the location of the electron Fermi level at the CdS/absorber interface when measured from the conduction band edge and from the valence band edge of the absorber, respectively. The quantity ϕ_n specifies the position of the interface neutrality level with respect to the valence band edge. The potential drop across the CdS buffer of thickness d_b is illustrated by V_D^b and the band bending over the space charge width w_a of the absorber is indicated by V_D^a. The conduction band offsets between the window and the buffer and between the absorber and the buffer are ΔE_C^{wb} and ΔE_C^{ab}, respectively. The surface defect layer lowers the valence band edge of the absorber towards the active interface and introduces an internal valence band offset ΔE_V^i.

First, we write the charge neutrality condition for the junction

$$Q_n + qd_bN_b + qN_I(\Delta E_{Fn} + \phi_n - E_g - \Delta E_V^i) - qN_aw_a = 0. \qquad (6.1)$$

Here, Q_n are the charges in the ZnO window layer, q is the elementary charge, ΔE_{Fn} is the energy distance between the electron Fermi level and the conduction band edge of the absorber at the junction interface, N_I is the density of states per unit area and unit energy at this heterointerface, and ϕ_n is the position of the interface neutrality level with respect to the valence band edge [33]. The four terms in Eq. 6.1 describe the charges in the ZnO window, in the CdS buffer, at the CdS/absorber interface, and in the absorber, respectively.

We obtain the second equation from the energy sum in the CdS/ZnO side of the band diagram, assuming a flat electron Fermi level across the CdS buffer layer [33]. Accordingly, the sum of the conduction band offsets $\Delta E_{\mathrm{C}}^{\mathrm{ab}}$ between the absorber and the buffer (taken as negative in Fig. 6.2), $\Delta E_{\mathrm{C}}^{\mathrm{wb}}$ between the window and the buffer, and the potential drop across the buffer layer $V_{\mathrm{D}}^{\mathrm{b}} = q N_{\mathrm{b}} d_{\mathrm{b}}^2 / (2\varepsilon_{\mathrm{b}}) + Q_n d_{\mathrm{b}}/\varepsilon_{\mathrm{b}}$ must equal the quantity ΔE_{Fn}

$$\Delta E_{\mathrm{C}}^{\mathrm{wb}} - \Delta E_{\mathrm{C}}^{\mathrm{ab}} + \frac{q N_{\mathrm{b}}}{2\varepsilon_{\mathrm{b}}} d_{\mathrm{b}}^2 + Q_n \frac{d_{\mathrm{b}}}{\varepsilon_{\mathrm{b}}} - \Delta E_{\mathrm{Fn}} = 0, \tag{6.2}$$

where ε_{b} is the dielectric constant of the buffer material. Here and in the following, all energies are expressed in units of the electrostatic potential.

The energy sum within the absorber part of the biased heterojunction yields the third equation [33]. The bandgap energy E_{g} in the bulk of the absorber equals the sum of the energy distance ΔE_{Fn}, the applied voltage V, the difference ζ between the hole Fermi energy and the valence band edge of the chalcopyrite in the quasineutral region, and the band bending of the absorber $V_{\mathrm{D}}^{\mathrm{a}} = q N_{\mathrm{a}} w_{\mathrm{a}}^2 / (2\varepsilon_{\mathrm{a}})$ (written in the depletion approximation). Therefore, we have

$$\Delta E_{\mathrm{Fn}} + V + \varsigma + \frac{q N_{\mathrm{a}}}{2\varepsilon_{\mathrm{a}}} w_{\mathrm{a}}^2 - E_{\mathrm{g}} = 0, \tag{6.3}$$

where ε_{a} is the dielectric constant of the absorber. At this point, we consider the recombination at the (inverted) CdS/absorber interface as dominant for the open-circuit voltage V_{OC} of the device. If the recombination centers are not very close to the conduction band, the recombination current j_{rec} is dominated by the concentration $p|_{\mathrm{if}}$ of the free holes at the interface[1] [1]

$$j_{\mathrm{rec}} = q S_{\mathrm{p}} \, p|_{\mathrm{if}} = q S_{\mathrm{p}} N_{\mathrm{V}} \exp\left(-\frac{V_{\mathrm{D}}^{\mathrm{a}} + \varsigma + \Delta E_{\mathrm{V}}^{\mathrm{i}}}{k_{\mathrm{B}} T}\right), \tag{6.4}$$

where N_{V} is the effective density of states in the valence band of the SDL, S_{p} is the interface recombination velocity for holes and $k_{\mathrm{B}} T$ is the thermal energy. At open circuit no current flows across the junction and the recombination current compensates the short-circuit current j_{sc} of the device [1]. Accordingly, we write the band bending $V_{\mathrm{D}}^{\mathrm{a}}|^{\mathrm{OC}}$ in the absorber under open-circuit conditions

$$V_{\mathrm{D}}^{\mathrm{a}}|^{\mathrm{OC}} = k_{\mathrm{B}} T \ln\left(\frac{q S_{\mathrm{p}} N_{\mathrm{V}}}{j_{\mathrm{sc}}}\right) - \zeta - \Delta E_{\mathrm{V}}^{\mathrm{i}}. \tag{6.5}$$

Equations 6.1, 6.2, and 6.3 give the voltage-dependent space charge width w_{a} in analogy to the equilibrium case [33]

---

[1] When the concentration of electrons at the CdS/absorber interface is non-negligible, the general Shockley–Read–Hall expression must be considered for the recombination rate.

$$w_{a} = \frac{\varepsilon_{a}}{\varepsilon_{b}} \frac{d_{b}}{1 + \nu_{I}} + \sqrt{\left(\frac{\varepsilon_{a}}{\varepsilon_{b}} \frac{d_{b}}{1 + \nu_{I}}\right)^{2} + \frac{2\varepsilon_{a}}{qN_{a}} \left(\frac{E_{1} + \nu_{I}E_{2}}{1 + \nu_{I}} - V\right)}, \qquad (6.6)$$

where $E_{1} \equiv E_{g} - \varsigma - \Delta E_{C}^{wb} + \Delta E_{C}^{ab} + qN_{b}d_{b}^{2}/(2\varepsilon_{b})$, $E_{2} \equiv \phi_{n} - \varsigma - \Delta E_{V}^{i}$, and $\nu_{I} = qN_{I}d_{b}/\varepsilon_{b}$ is the normalized density of interface states. Resolving Eq. 6.6 for V and using the depletion approximation for the absorber yields the open-circuit voltage

$$V_{OC} = \frac{E_{1} + \nu_{I}E_{2}}{1 + \nu_{I}} - V_{D}^{a}\big|^{OC} - \frac{1}{1 + \nu_{I}} \frac{d_{b}}{\varepsilon_{b}} \sqrt{2q\varepsilon_{a} N_{a} V_{D}^{a}\big|^{OC}}, \qquad (6.7)$$

where the band bending $V_{D}^{a}\big|^{OC}$ is given by Eq. 6.5.

As it follows from Eqs. 6.3 and 6.5, the interface barrier $\Phi_{b}^{p} = E_{g} + \Delta E_{V}^{i} - \Delta E_{Fn}$ defined in Fig. 6.2 as the energy distance of the electron Fermi level to the valence band edge of the absorber at the junction interface, may be written under open-circuit conditions as

$$\Phi_{b}^{p}\big|^{OC} = V^{OC} + k_{B}T \ln\left(\frac{qS_{p}N_{V}}{j_{sc}}\right). \qquad (6.8)$$

The above analytical model allows one to compute the open-circuit voltage V_{OC} as a function of the temperature T, the material parameters E_{g}, N_{a}, N_{b}, ε_{a}, ε_{b}, ς, the geometry parameter d_{b}, the interface conditions defined by ϕ_{n}, N_{I}, S_{p}, ΔE_{C}^{ab}, ΔE_{C}^{wb}, and the short-circuit current j_{sc} of the device.

Figure 6.3 displays the analytically calculated dependence of the open-circuit voltage V_{OC} at 300 K on the interface state density N_{I} for (a) a low-gap chalcopyrite device with a flat conduction band alignment at the CdS/absorber interface and (b) a wide-gap device with a conduction band cliff of -0.55 eV at the respective interface.

We have assumed different values of the charge neutrality level ϕ_{n} and of the internal valence band offset ΔE_{V}^{i}. The values of the open-circuit interface barrier $\Phi_{b}^{p}\big|^{OC}$ corresponding to the analytical calculations are also displayed on the right vertical axis, which is derived from the left axis using Eq. 6.8. Additional numerical calculations using SCAPS-1D [36] as a device simulator are included. The numerical modeling is carried out using a ZnO/CdS/SDL/chalcopyrite heterostructure (without an *intrinsic* ZnO) where the doped ZnO window has a high n-type doping density of 10^{20} cm^{-3} in order to approach the situation described in Fig. 6.2. While bulk defects are removed from all the semiconductor partners, a deep neutral defect is designed as dominant recombination trap (not contributing to the pinning) at the SDL/CdS interface. In addition, energetically uniform distributions of donors and acceptors both centered at the charge neutrality level are assumed as defects that pin the Fermi level at the active junction interface. In this way the recombination and the pinning mechanisms are physically disconnected.

Figure 6.3 shows that densities of interface states of the order of 10^{13} cm^{-2} eV^{-1} are sufficient to pin the Fermi level at the CdS/absorber interface at the

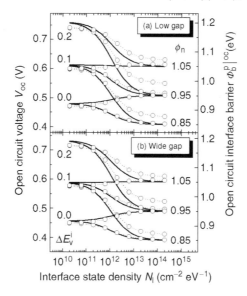

Fig. 6.3. Analytical (*lines*) and numerical (*symbols*) calculations of the opencircuit voltage V_{OC} at 300 K as a function of interface state density N_I for (**a**) a lowgap chalcopyrite device with $E_g = 1.15\,\text{eV}$, $\Delta E_C^{ab} = 0\,\text{eV}$, and (**b**) a widegap chalcopyrite device with $E_g = 1.7\,\text{eV}$, $\Delta E_C^{ab} = -0.55\,\text{eV}$. The valence band edge of the absorber is set at the same energy position for both devices. Different assumed values (in eV) of the internal valence band offset ΔE_V^i and of the charge neutrality level ϕ_n are indicated. The doping densities and the dielectric constants of the buffer (thickness $d_b = 50\,\text{nm}$) and absorber (effective density of states in the valence band $N_V = 1.5 \times 10^{19}\,\text{cm}^3$) are $N_b = N_a = 10^{16}\,\text{cm}^3$ and $\varepsilon_b = \varepsilon_a = 10\varepsilon_0$, where ε_0 is the vacuum permittivity. The conduction band offset between the window and the buffer is $\Delta E_C^{wb} = 0.1\,\text{eV}$ and the recombination velocity for holes at the CdS/absorber interface is $S_p = 4.2 \times 10^5\,\text{cm/s}$. Shortcircuit currents j_{SC} of 38.3 mA/cm^2 for the low gap and 19.8 mA/cm^2 for the widegap device are used in the analytical calculations. The surface defect layer has a thickness of 50 nm in numerical calculations and, except E_g, has the same properties as the absorber. The right vertical axis illustrates the linear relationship between the open circuit interface barrier $\Phi_b^p|_{OC}$ and V_{OC} according to Eq. 6.8

neutrality level [33], which directly influences, in this case, the open-circuit voltage of the device. Under decreasing density of interface states, the Fermi level is gradually unpinned. The surface defect layer with a larger bandgap than the absorber bulk increases the open-circuit voltage at low interface state densities when the Fermi level at the interface is unpinned. The presence of the surface defect layer in Cu-poor chalcopyrite heterojunctions has a beneficial effect upon the electrical transport [22].

The numerical modeling fairly well reproduces the analytical trends in Fig. 6.3 when using essentially the same set of parameters. We mention, however, that a poorer agreement occurs when the charge neutrality level is closer to the conduction band edge or to midgap. While the calculations with $\Delta E_V^i = 0.0\,\text{eV}$, $\phi_n = 1.05\,\text{eV}$, and those with $\Delta E_V^i = 0.2\,\text{eV}$, $\phi_n = 0.85\,\text{eV}$ are not shown in Fig. 6.3, a considerable disagreement, especially for the latter set of data arises between analytical and numerical investigations at high interface state densities. One should consider the discrepancies that arise when approaching a real heterojunction device in numerical simulations.

The open-circuit voltage limitation by interface recombination is approximately similar for both devices in Fig. 6.3. The differences in the open-circuit voltage are mainly determined by the short-circuit current j_{sc} (otherwise identical parameters except the absorber bandgap and the conduction band offset at the active junction interface are assumed for both heterostructures), which is considerably lower for the wide-gap device than for the low-gap counterpart. We note that the open-circuit voltage is nearly independent of the bandgap energy of the absorber as long as the recombination at the CdS/absorber interface is dominant and the valence band edge of the absorber remains at the same energy position.

6.5 Numerical Simulations

We now turn from the idealized situation, which allows for an analytical approach, to a more specific study on the influence of shallow and deep defects within the SDL, where we completely rely on numerical simulations [35]. We consider the *doping model*, in which the surface layer is an n-type doped material and the *pinning model*, which views the surface layer as a material that is type inverted due to Fermi-level pinning by donor-like defects at the interface with CdS. We use numerical device simulations to study the implications of these two scenarios on the performance of the solar cell. The internal valence band offset ΔE_V^i between the surface defect layer and the bulk of the absorber is the primary ingredient that is necessary to obtain high-efficiency devices as can be concluded, e.g., from earlier numerical device simulations [37]. The advantage of the bandgap widening toward the CIGS surface is the reduction of recombination at the interface toward the CdS buffer layer [37,38] (see also Sect. 6.4). In the following, we will show that this finding is independent of the specific approach (doping/pinning model), though there exist differences between the two approaches in some details.

6.5.1 Numerical Modeling

As in Sect. 6.4, we perform numerical modeling using the device simulator SCAPS-1D [36]. However, there are some differences between the parameters used in the numerical modeling in Sect. 6.4 and the parameters used for simulation in Sect. 6.5 (the latter set of parameters is described in the following). In this section, the device structure consists of a doped and intrinsic ZnO

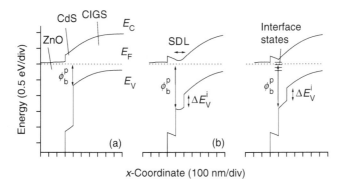

Fig. 6.4. Equilibrium band diagrams of ZnO/CdS/Cu(In,Ga)Se$_2$ heterostructures (**a**) without a surface defect layer SDL, (**b**) with an SDL assumed as n-type material with doping concentration 10^{17} cm^{-3}, and (**c**) with an SDL that is type inverted due to Fermi-level pinning by interface donors of density 10^{13} cm^2/eV. The quantity ΔE_V^i denotes the valence band offset between SDL and Cu(In,Ga)Se$_2$, and Φ_b^p shows the energy distance between Fermi level E_F and valence band edge E_V at the CdS/absorber interface

window, the CdS buffer, the 50 nm-thick surface defect layer SDL, and the chalcopyrite CIGS absorber. Figure 6.4 depicts the band diagram of a chalcopyrite heterojunction (a) without surface defect layer, (b) with an SDL that is assumed to consist of n-type material (*doping model*), and (c) with an SDL that is type inverted due to Fermi-level pinning by donors at the CdS/SDL interface (*pinning model*).

In the latter case, a density of interface states of the order of 10^{13} cm^{-2} eV^{-1} is sufficient to pin the Fermi level at the CdS/SDL interface at the charge neutrality level [33]. We consider in the *doping* approach a deep neutral defect as an effective recombination center at the CdS/SDL interface. In the *pinning* approach, we use (in addition to the neutral, recombination active defect) a donor-like surface defect at the CdS/SDL interface (with $E_c - E_t = 0.2$ eV) for Fermi-level pinning. Defects in the bulk of the CIGS, SDL, CdS, and intrinsic ZnO are also employed for bulk recombination. Table 6.1 lists the main parameters of the simulation used in the following.

The starting point of the simulations is a CIGS absorber material (bandgap energy $E_g = 1.15$ eV), without SDL according to the band diagram shown in Fig. 6.4(a) that could yield a device with efficiency $\eta = 19.3\%$ and open-circuit voltage $V_{OC} = 682$ mV. However, introducing recombination (with an interface recombination velocity $S = 10^6$ cm/s) at the absorber buffer interface immediately leads to $\eta = 7.2\%$ and $V_{OC} = 328$ mV. Starting from that unfavorable situation, Sect. 6.5.2 investigates the reduction of interface recombination by the presence of the SDL.

Table 6.1. Parameters of different semiconductor layers used in the simulation

	CIGS	SDL	CdS	i-ZnO	doped-ZnO
thickness (µm)	1	0.05	0.05	0.08	0.2
bandgap (eV)	E_g	$E_g + \Delta E_V^i$	2.42	3.3	3.3
El. affinity (eV)	4.5	4.5	4.5	4.7	4.7
ε	10	10	10	10	10
N_C (cm^{-3})	6.7×10^{17}	6.7×10^{17}	2×10^{19}	5×10^{18}	5×10^{18}
N_V (cm^{-3})	1.5×10^{19}	1.5×10^{19}	2×10^{19}	5×10^{18}	5×10^{18}
$\nu_{th,n}$ (cm/s)	3.9×10^7	3.9×10^7	1×10^7	1×10^8	1×10^8
$\nu_{th,p}$ (cm/s)	1.4×10^7	1.4×10^7	1×10^7	1×10^8	1×10^8
μ_n (cm^2/V/s)	100	100	0.1	1	1
μ_p (cm^2/V/s)	10	10	5	1	1
doping (cm^{-3})	1×10^{16} (p)	1×10^{16} (p)	8×10^{15} (n)	5×10^{17} (n)	1×10^{18} (n)
		1×10^{17} (n)			

The quantity ε denotes the relative dielectric permittivity, N_C and N_V are the conduction and valence band effective density of states, $\nu_{th,n}$ and $\nu_{th,p}$ are the thermal velocities for electrons and holes, and μ_n and μ_p the mobilities

6.5.2 Influence of Surface Bandgap Widening

Figure 6.5 shows the calculated efficiency η and open-circuit voltage V_{OC} of the device as a function of doping concentration in the SDL (doping model, Fig. 6.5 a and b), and as a function of density of interface states at the CdS/SDL interface (pinning model, Fig 6.5c and d). The internal valence band offset ΔE_V^i between the SDL and the chalcopyrite material is taken as a parameter and an additional bandgap profile (half symbols) with a gradual lowering of the valence band edge of the absorber toward the interface with CdS is also considered.

Without bandgap widening (open symbols in Fig. 6.5), high n-type bulk doping concentrations of 10^{17} cm^{-3} in the SDL (a and b), or Fermi-level pinning at the interface with CdS (c and d) with interface state density of around 10^{13} cm^{-2} eV^{-1} are necessary to improve the calculated efficiency η and the open-circuit voltage V_{OC} of the device. The internal valence band offset $\Delta E_V^i = 0.2$ eV (full symbols) or the graded bandgap profile (half symbols) makes the calculated η and V_{OC} of the device more independent of the n-type doping in the SDL (in the doping model), or of the density of interface states at the CdS/SDL interface (in the pinning model). This is because the bandgap widening toward the absorber surface increases the energy barrier Φ_b^p at the CdS/SDL interface and therefore diminishes the interface recombination [37,38] (see also Sect. 6.4). The recombination in the SCR of the absorber thus becomes more prominent, a situation characteristic to devices based on Cu-poor chalcopyrite absorbers [19,39]. In turn, devices based on absorbers prepared under Cu-rich conditions show relatively low interface barriers and are therefore assumed to be dominated by recombination at the CdS/absorber interface [22].

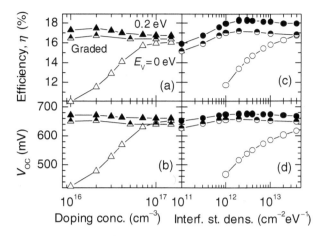

Fig. 6.5. Calculated efficiency η and open-circuit voltage V_{OC} for the n-type SDL (doping model, **a** and **b**), and for the SDL with Fermi-level pinning (**c** and **d**). An internal valence band offset $\Delta E_{\mathrm{V}}^{\mathrm{i}} = 0.2\,\mathrm{eV}$ (*filled symbols*) or a graded structure (*half symbols*) makes the parameters, to a first order, independent of the doping concentration in the SDL (within the doping model) or of the density of states at the CdS/SDL interface (within the pinning model). The devices with no internal band offset (*open symbols*) require high n-type doping or Fermi-level pinning

6.5.3 Influence of Excess Defects

This subsection investigates the effect of excess recombination centers in the bulk of the SDL as well as at the SDL/absorber interface. Such defects have to be taken into account as, for instance, recent investigations with high-resolution transmission electron microscopy unveil a large number of structural defects in this specific region [40]. For our simulations, we have assumed midgap defects ($E_{\mathrm{t}} - E_{\mathrm{V}} = 0.7\,\mathrm{eV}$, $\sigma_{\mathrm{n}} = \sigma_{\mathrm{p}} = 10^{-13}\,\mathrm{cm}^{-2}$) in the SDL reducing the hole diffusion length L_{p} in the defect layer to 4.3 nm at a defect concentration of $10^{18}\,\mathrm{cm}^{-3}$. Figure 6.6(a) shows that the efficiency η of a device with Fermi-level pinning, though starting at a slightly higher efficiency level, is more sensitive to excess defects than a device that is surface type inverted by an n-type doped SDL. However, due to the internal band offset $\Delta E_{\mathrm{V}}^{\mathrm{i}} = 0.2\,\mathrm{eV}$; both types of devices can accommodate a large number of defects without significant losses in V_{OC} (Fig. 6.6b). The reduction of efficiency with increasing bulk defect concentration results from the loss in short-circuit current density j_{sc} (not shown in Fig. 6.6).

Excess defects located at the SDL/CIGS interface are also important as possible recombination traps at this internal interface. Therefore, we study defects at the internal SDL/CIGS interface ($E_{\mathrm{t}} - E_{\mathrm{V}} = 0.6\,\mathrm{eV}$, $\sigma_{\mathrm{n}} = \sigma_{\mathrm{p}} = 10^{-15}\,\mathrm{cm}^{-2}$, leading to an interface recombination velocity of $S = 10^5\,\mathrm{cm/s}$ at a density $10^{13}\,\mathrm{cm}^{-2}\,\mathrm{eV}^{-1}$). Figure 6.6(c) and (d) shows calculated efficiencies

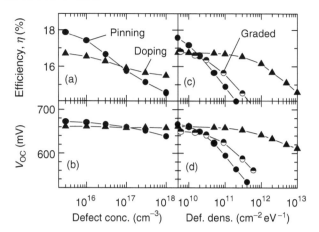

Fig. 6.6. Efficiencies η (**a**) and open-circuit voltages V_{OC} (**b**) calculated for increasing density of bulk defects in the SDL. The SDL is assumed to be n-type (*triangles*) or inverted by Fermi-level pinning (*circles*). An internal valence band offset $\Delta E_V^i = 0.2\,eV$ is assumed in both cases. Parts (**c**) and (**d**) show the calculated efficiencies and open-circuit voltages as a function of density of interface states at the internal SDL/CIGS interface. *Triangles* refer to the doping model, *full circles* to the pinning approach and *half circles* refer to a graded bandgap with pinning

and the open-circuit voltages for the doping model (triangles) and the pinning model (full circles). We have also considered a graded surface bandgap with pinning (half circles) that is simulated by three 16 nm-thick layers with a gradually increasing bandgap. One-third of the defects are put at each of the three interfaces. From Fig. 6.6(c) and (d), we learn that the doping approach provides a markedly higher tolerance to internal interface recombination, when compared to the pinning model or the graded band-gap device.

Thus, we conclude from our numerical results that the surface bandgap widening in CIGS photovoltaic absorbers provides a certain tolerance to excess defects within this specific close-to-surface region. Note that *without* surface bandgap widening, i.e., $\Delta E_V^i = 0$, the device performance drops from 16% to 13% for the doping model and even from 18% to 11% in the pinning model when turning from zero to 10^{18} cm^{-3} bulk defect density in the SDL. The corresponding decrease of efficiency with $\Delta E_V^i = 0.2\,eV$, as shown in Fig. 6.6(a), is from 16.8% to 15.5% (in the doping model) and from 18% to 14.7% (in the pinning model). When comparing the doping model with the pinning model, we find that the tolerance of an n-type doped surface layer with respect to defects in its bulk or at its internal interface with the absorber is considerably larger than that of devices where surface-type inversion is warranted by surface states.

6.5.4 Wide-Gap Chalcopyrite Absorbers

This section focuses on wide-gap chalcopyrite absorbers with a band energy of $E_g = 1.65\,\mathrm{eV}$ (CuGaSe$_2$). Here it is assumed that the increase of E_g with respect to the low-gap material follows from an up-shift of the conduction band energy. To ensure the dominance of interface recombination, we have assumed the same good electronic quality, i.e., low bulk defect density, as in the case of the low-gap absorbers in the preceding sections.

Figure 6.7(a) shows the calculated efficiencies η and fill factors FF for the respective ZnO/CdS/SDL/chalcopyrite device as a function of the internal valence band offset ΔE_V^i between the SDL and the chalcopyrite. Again, we have considered both the doping model and the pinning approach. The efficiency η increases in a similar manner within both models upon increasing the internal valence band offset ΔE_V^i in the investigated range. This is because of the suppression of interface recombination by the surface bandgap enhancement. However, even with the most optimistic assumptions concerning bulk properties and surface bandgap widening, efficiencies of the order of 18% cannot be obtained with the wide-gap material.

As can be seen from Fig. 6.7(a), the fill factors (FF) remain at relatively good values only at low internal valence band offsets ΔE_V^i, while at larger

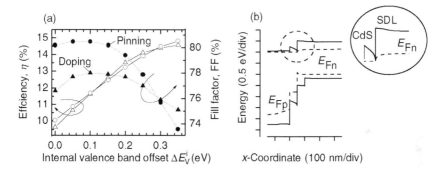

Fig. 6.7. (a) Calculated efficiency η (*open symbols*) and fill factor FF (*filled symbols*) of the wide-gap chalcopyrite device as a function of the internal valence band offset ΔE_V^i between the SDL and the chalcopyrite. *Open and filled triangles* refer to the doping model, the *open and filled circles* are calculated in the pinning approach. The efficiencies increase with increasing the internal valence band offset and the fill factors deteriorate at higher ΔE_V^i. (b) The band diagram according to the pinning model with $\Delta E_V^i = 0.35\,\mathrm{eV}$ calculated for a device under illumination close to the maximum power point ($V = 1.0$ V). The external voltage V exceeds the zero-bias built-in voltage leading to a barrier that builds up under working conditions. The fill factor losses (**a**) result from a drop of the electron Fermi level E_{Fn} across the SDL/CdS interface (*inset*)

ΔE_V^i the fill factors deteriorate and finally put an upper limit to the efficiency of the wide-gap device. Thus, even if one could improve the relatively poor bulk properties of wide-gap chalcopyrites [39, 41], the unfavorable band alignment (cf. Fig. 6.7b) at the SDL/CdS/ZnO interfaces limits carrier collection under working conditions and, as a consequence, the FF of wide-gap chalcopyrite devices.

6.6 Conclusion

For any heterojunction solar cell, recombination at the heterointerface is a major challenge. This statement holds especially for nonepitaxial, nonlattice-matched interfaces like that of the CdS/Cu(In,Ga)(Se,S)$_2$ solar cell. Therefore, the fact that most experimental investigations, discussed in Sect. 6.2, point to bulk instead of interface recombination as the dominant recombination path in this device is not straightforward and requires a special explanation. The present contribution has shown, by the analytical model in Sect. 6.4 as well as by the more refined numerical simulations in Sect. 6.5, that the primary quality of the Cu(In,Ga)(Se,S)$_2$ surface, which becomes the photovoltaic active interface in the finished device, is the Cu-poor surface defect layer. The surface bandgap widening of this surface layer suppresses interface recombination. Therefore, bulk recombination dominates in these devices, a situation that is more favorable as it allows for a much higher open-circuit voltage than that we expect from devices that are dominated by interface recombination. However, the open-circuit voltage allowed by interface recombination is almost independent of the bandgap energy of the chalcopyrite when the valence band edge of the absorber remains at the same energy position. Though the beneficial effect of the surface defect layer is retained, interface recombination becomes increasingly important when turning to wide-gap chalcopyrite absorbers. This is why experimental results from CuGaSe$_2$ solar cells exhibit ambiguous results for what concerns the dominating recombination path.

The numerical simulations of Sect. 6.5 are used to refine our investigations of the role of the surface bandgap widening. We have considered the SDL either as n-type material or as a material that is type inverted due to Fermi-level pinning by defects at the CdS/SDL interface. Without band gap widening, high n-type doping concentrations in the SDL or Fermi-level pinning at the CdS/SDL interface are necessary for high-efficiency devices. The bandgap widening toward the CIGS surface tends to maintain an improved device efficiency and open-circuit voltage under decreasing n-type doping concentrations in the SDL or under decreasing densities of states at the CdS/SDL interface. Because of its larger bandgap, the SDL can accommodate a relatively high number of excess defects in its bulk or at internal interfaces. Here, an n-type SDL appears more suitable to preserve the efficiency under increasing numbers of excess defects than an SDL with Fermi-level pinning. However, the *primary* quality of the SDL—irrespective of the specific assumptions on its doping type – consists in the internal valence band offset that suppresses recombination at the interface toward the CdS buffer layer.

Calculations for wide-gap chalcopyrite absorbers show that also in such devices the internal valence band offset implies an increase of device efficiency. However, the available efficiencies range only up to 15.5% even when assuming the same bulk quality that leads to $\eta > 18\%$ for low-gap absorbers. The limitations result, on the one hand, from the fact that interface recombination does not decrease with increasing bandgap energy E_g as long as E_g increases because of an upward shift of the conduction band energy E_C. Additionally, the unfavorable band alignment at the wide-gap chalcopyrite/CdS/ZnO interfaces lowers the fill factor and decreases the device performance further. Though the best wide-gap chalcopyrite solar cells, e.g., based on CuGaSe$_2$, may still be limited by the poor bulk quality of the absorber material, considerable improvements would require not only better absorber quality but also alternative buffer/window combinations.

Acknowledgments

The authors thank T. Schlenker, W. Brendle, and J. Mattheis for fruitful collaboration. We also thank H.W. Schock and J.H. Werner for valuable support. This work was funded by the German Ministry for Education and Research under contract no. 01SF0023 (Hochspannungsnetz).

References

1. Rau, U., Schock, H.W.: Cu(In,Ga)Se$_2$ solar cells. In: Archer, M.D., Hill, R. (eds.) Clean Electricity from Photovoltaics, pp. 277–292. Imperial College Press, London (2001)
2. Dimmler, B., Schock, H.W.: Scalability and pilot operation in solar cells of CuInSe$_2$ and their alloys. Prog. Photovolt.: Res. Appl. 6, 193–199 (1998)
3. Jasenek, A., Rau, U.: Defect generation in Cu(In,Ga)Se$_2$ heterojunction solar cells by high-energy electron and proton irradiation. J. Appl. Phys. 90, 650–658 (2001)
4. Guillemoles, J.F.: Stability of Cu(In,Ga)Se$_2$ solar cells: a thermodynamic approach. Thin Solid Films 361–362, 338–345 (2000)
5. Contreras, M.A., Egaas, B., Ramanathan, K., Hiltner, J., Swartzlander, A., Hasoon, F., Noufi, R.: Progress toward 20% efficiency in Cu(In,Ga)Se$_2$ polycrystalline thin-film solar cells. Prog. Photovolt.: Res. Appl. 7, 311–316 (1999)
6. Herberholz, R., Nadenau, V., Rühle, U., Köble, C., Schock, H.W., Dimmler, B.: Prospects of wide-gap chalcopyrites for thin film photovoltaic modules. Sol. Energy Mater. Sol. Cells 49, 227–237 (1997)
7. Siebentritt, S.: Wide gap chalcopyrites: material properties and solar cells. Thin Solid Films 403–404, 1–8 (2002)

8. Rau, U.: Electronic properties of wide-gap Cu(In,Ga)(Se,S)$_2$ solar cells. In: Kurosawa, K., Kazmerski, L.L., McNellis, B., Yamaguchi, M., Wronski, C., Sinke, W.C. (eds.) Proceedings of the 3rd World Conference on Photovoltaic Energy Convention, Osaka, pp. 2847–2852, (2003)
9. Sites, J.R.: Separation of voltage loss mechanisms in polycrystalline solar cells. In: Conference Record of the 20th IEEE Photovoltaic Specialists Conference. IEEE, New York, pp. 1604–1607 (1988)
10. Roy, M., Damaskinos, S., Phillips, J.E.: The diode current mechanism in CuInSe$_2$/(CdZn)S heterojunctions. In: Conference Record of the 20th IEEE Photovoltaic Specialists Conference. IEEE, New York, pp. 1618–1623 (1988)
11. Shafarman, W.N., Phillips, J.E.: Diode analysis of CuInSe$_2$ solar cells. In: Conference Record of the 22nd IEEE Photovoltaic Specialists Conference. IEEE, New York, pp. 934–939 (1991)
12. Santamaria, J., Martil, I., Iborra, E., Diaz, G.G., Quesada, F.S.: Electrical characterization of all-sputtered CdS/CuInSe$_2$ solar cell heterojunctions. Solar Cells 28, 31–39 (1990)
13. Phillips, J.E., Birkmire, R.W., McCandless, B.E., Meyers, P.V., Shafarman, W.N.: Polycrystalline heterojunction solar cells: a device perspective. Phys. Stat. Sol. B 194, 31–39 (1996)
14. Mitchell, K.W., Liu, H.I.: Device analysis of CuInSe$_2$ solar cells. In: Conference Record of the 20th IEEE Photovoltaic Specialists Conference. IEEE, New York, pp. 1461–1468 (1988)
15. Yoo, J.B., Fahrenbruch, A.L., Bube, R.H.: Transport mechanisms in ZnO/CdS/CuInSe$_2$ solar cells. J. Appl. Phys. 68, 4694–4699 (1990)
16. Walter, T., Herberholz, R., Schock, H.W.: Distribution of defects in polycrystalline chalcopyrite thin films. Solid State Phenom. 51/52, 309–316 (1996)
17. Rau, U.: Tunneling-enhanced recombination in Cu(In,Ga)Se$_2$ heterojunction solar cells. Appl. Phys. Lett. 74, 111–113 (1999)
18. Malmström, J., Wennerberg, J., Bodegård, M., Stolt, L.: Influence of Ga on the current transport in Cu(In,Ga)Se$_2$ thin film solar cells. In: Proceedings of the 17th European Photovoltaic Solar Energy Conference. WIP—Munich and ETA—Florence, pp. 1265–1268, (2001)
19. Rau, U., Schmidt, M., Jasenek, A., Hanna, G., Schock, H.W.: Electrical characterization of Cu(In,Ga)Se$_2$ thin-film solar cells and the role of defects for the device performance. Sol. Energy Mater. Sol. Cells 67, 137–143 (2001)
20. Nadenau, V., Rau, U., Jasenek, A., Schock, H.W.: Electronic properties of CuGaSe$_2$-based heterojunction solar cells. Part I. Transport analysis. J. Appl. Phys. 87, 584–593 (2000)
21. Saad, M., Kassis, A.: Analysis of illumination-intensity-dependent j–V characteristics of ZnO/CdS/CuGaSe$_2$ single crystal solar cells. Sol. Energy Mater. Sol. Cells 77, 415–422 (2003)
22. Turcu, M., Pakma, O., Rau, U.: Interdependence of absorber composition and recombination mechanism in Cu(In,Ga)(Se,S)$_2$ heterojunction solar cells. Appl. Phys. Lett. 80, 2598–2600 (2002)
23. Turcu, M., Rau, U.: Fermi-level pinning at CdS/Cu(In,Ga)(Se,S)$_2$ interfaces: effect of chalcopyrite alloy composition. J. Phys. Chem. Solids 64, 1591–1595 (2003)
24. Reiß, J., Malmström, J., Werner, A., Hengel, I., Klenk, R., Lux-Steiner, M.C.: Current transport in CuInS$_2$ solar cells depending on absorber preparation. Mater. Res. Soc. Symp. Proc. 668, H9.4.1–H9.4.6 (2001)

25. Hengel, I., Neisser, A., Klenk, R., Lux-Steiner, M.C.: Current transport in CuInS$_2$:Ga/CdS/ZnO-solar cells. Thin Solid Films 361–362, 458–462 (2000)
26. Klenk, R.: Characterisation and modelling of chalcopyrite solar cells. Thin Solid Films 387, 135–140 (2001)
27. Schmid, D., Ruckh, M., Grunwald, F., Schock, H.W.: Chalcopyrite/defect chalcopyrite heterojunctions on the basis of CuInSe$_2$. J. Appl. Phys. 73, 2902–2909 (1993)
28. Morkel, M., Weinhardt, L., Lohmüller, B., Heske, C., Umbach, E., Riedl, W., Zweigart, S., Karg, F.: Flat conduction-band alignment at the CdS/CuInSe$_2$ thin-film solar-cell heterojunction. Appl. Phys. Lett. 79, 4482–4484 (2001)
29. Schmid, D., Ruckh, M., Schock, H.W.: Photoemission studies on Cu(In,Ga)Se$_2$ thin films and related binary selenides. Appl, Surf. Sci. 103, 409–421 (1996)
30. Ramanathan, K., Noufi, R., Granata, J., Webb, J., Keane, J.: Prospects for in situ junction formation in CuInSe$_2$ based solar cells. Sol. Energy Mater. Sol. Cells 55, 15–22 (1998)
31. Sugiyama, T., Chaisitsak, S., Yamada, A., Konagai, M., Kudriavtsev, Y., Godines, A., Villegas, A., Asomoza, R.: Formation of pn homojunction in Cu(InGa)Se-2 thin film solar cells by Zn doping. Jpn. J. Appl. Phys., Part 1 39, 4816 (2000)
32. Jiang, C.S., Hasoon, F.S., Moutinho, H.R., Al-Thani, H.A., Romero, M.J., Al-Jassim, M.M.: Direct evidence of a buried homojunction in Cu(In,Ga)Se-2 solar cells. Appl. Phys. Lett. 82, 127–129 (2003)
33. Herberholz, R., Rau, U., Schock, H.W., Haalboom, T., Gödecke, T., Ernst, F., Beilharz, C., Benz, K.W., Cahen, D.: Phase segregation, Cu migration and junction formation in Cu(In,Ga)Se$_2$. Eur. Phys. J. AP 6, 131–139 (1999)
34. Rau, U., Braunger, D., Herberholz, R., Schock, H.W., Guillemoles, J.F., Kronik, L., Cahen, D.: Oxygenation and air-annealing effects on the electronic properties of Cu(In,Ga)Se$_2$ films and devices. J. Appl. Phys. 86, 497–505 (1999)
35. Rau, U., Turcu, M.: Role of surface band gap widening in Cu(In,Ga)(Se,S)$_2$ thin-films for the photovoltaic performance of ZnO/CdS/Cu(In,Ga)(Se,S)$_2$ heterojunction solar cells. Mater. Res. Soc. Symp. Proc. 763, 335–340 (2003)
36. Niemegeers, A., Burgelman, M.: Numerical modelling of ac-characteristics of CdTe and CIS solar cells. In: Conference Record of the 25th IEEE Photovoltaic Specialists Conference. IEEE, New York, pp. 901–904 (1996)
37. Huang, C.H., Li, S.S., Anderson, T.J.: Device modeling and simulation of CIS-based solar cells. In: Conference Record of the 29th IEEE Photovoltaic Specialists Conference. IEEE, New York, pp. 748–751 (2002)
38. Dullweber, T., Hanna, G., Rau, U., Schock, H.W.: A new approach to high-efficiency solar cells by band gap grading in Cu(In,Ga)Se$_2$ chalcopyrite semiconductors. Sol. Energy Mater. Sol. Cells 67, 145–150 (2001)
39. Hanna, G., Jasenek, A., Rau, U., Schock, H.W.: Influence of the Ga-content on the bulk defect densities of Cu(In,Ga)Se$_2$. Thin Solid Films 387, 71–73 (2001)
40. Yan, Y., Jones, K.M., AbuShama, J., Al-Jassim, M.M., Noufi, R.: A TEM study of the microstructure evolution of Cu(In,Ga)Se$_2$ films from Cu-rich to In-rich. Mater. Res. Soc. Symp. Proc. 668, H6.10.1–H6.10.6 (2001)
41. Jasenek, A., Rau, U., Nadenau, V., Schock, H.W.: Electronic properties of CuGaSe$_2$-based heterojunction solar cells. Part II. Defect spectroscopy. J. Appl. Phys. 87, 594–602 (2000)

Shallow Defects in the Wide Gap Chalcopyrite CuGaSe$_2$

S. Siebentritt

Shallow defects in semiconductors are responsible for their doping behavior. This chapter investigates the doping effects of the shallow defects in CuGaSe$_2$. A critical review of the experimental methods photoluminescence (PL) and Hall investigations as well as a review of the available literature data on defects in CuGaSe$_2$ and CuInSe$_2$ are given. A comprehensive study of the shallow defects in CuGaSe$_2$ and CuInSe$_2$ for comparison by PL and Hall measurements is presented. Compositional dependencies are explicitly taken into account. It is found that CuGaSe$_2$ is dominated by a shallow donor and three acceptor states at 60, 100, and 150 meV from the valence band. The occurrence of the 150 meV defect is mostly independent of the composition while the 60 meV acceptor dominates at compositions close to and below the stoichiometry; the 100 meV acceptor dominates in materials grown under high Cu-excess. Self-compensation of native defects by native defects plays a major role in this material.

7.1 Introduction: Native Defects and Doping in CuGaSe$_2$

The amazing capability of all chalcopyrite materials to form over large deviations from stoichiometry is directly related to their doping behavior. Compositional variations from stoichiometry are made possible by the formation of native defects. The large tolerance to stoichiometry deviations can be seen best from the pseudobinary phase diagram, i.e., assuming that the Se concentrations do not deviate from the stoichiometric concentration. In CuGaSe$_2$, Ga concentrations of at least 28% are possible without losing the chalcopyrite structure [1]. These large Cu-deficiencies are only possible under the formation of Cu vacancies, Cu$_{Ga}$ antisites and their complexes. Thus, CuGaSe$_2$ is doped without the addition of intentional impurities. On the other hand, it is not possible to form Cu-rich CuGaSe$_2$ with a chalcopyrite structure. If grown under Cu-excess, stoichiometric CuGaSe$_2$ is formed with the Cu-excess precipitating in a secondary Cu–Se phase. The large tolerance to stoichiometry deviations is due to the low defect formation enthalpies in these materials compared to elemental semiconductors or the II–VI semiconductors. Typical defect formation enthalpies in Cu-chalcopyrites are around 1 ev as calculated by van Vechten's cavity model [2, 3] and by ab initio density functional theory [4, 5].

In a ternary system like CuGaSe$_2$, 12 native defects are conceivable: three vacancies, three interstitials and six antisite defects. Additionally, any complex of these could occur. Although the same native defects are possible in CuInSe$_2$ and CuGaSe$_2$ there is a fundamental difference in their doping behavior: CuInSe$_2$ can be p-type, when prepared under Se-excess or Cu-excess, or n-type, when prepared under Se-deficiency. On the contrary, CuGaSe$_2$ remains p-type under all stoichiometry deviations [6]. Therefore, it is necessary to investigate the shallow defects in CuGaSe$_2$ and CuInSe$_2$ to understand the differences in the doping behavior.

The question is whether there is a key to understand the different solar cell efficiencies of Cu-chalcopyrites with low bandgaps, i.e., Cu(In,Ga)Se$_2$ with low Ga content and those with high bandgaps, like CuGaSe$_2$. While the former have reached efficiencies close to 20% in the laboratory [7], the latter have not even yielded an efficiency of 10% up to now [8].

7.2 Basics: Photoluminescence and Electrical Transport

This section reviews briefly the basic theoretical concepts of PL and electrical transport in semiconductors. These models are needed to extract correct defect data from experimental results. There are certain misinterpretations reoccurring in the literature on PL and Hall measurements of chalcopyrites. These pitfalls will be point ed out here.

7.2.1 Defect Spectroscopy by Photoluminescence

There are a number of text books and reviews (see e.g., [9–14]) available on this topic, but none actually summarizes the models and experiments needed for the analysis of defects. Therefore, in this section, we will concentrate on facts and implications needed for the interpretation of the experimental data on defects discussed later.

Radiative Transitions in Semiconductors

After a semiconductor has been excited by light with energy above its bandgap energy, it returns to its equilibrium state by radiative and nonradiative processes. Of interest, here are only the radiative transitions. These include band-to-band transitions, excitonic transitions, transitions between a free carrier in the band and a defect state (free-to-bound transitions, FB), and transitions between defects (donor–acceptor-pair transitions, DA). Although it is

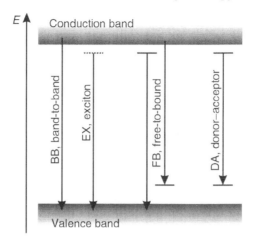

Fig. 7.1. Radiative processes in a semiconductor

possible to extract information on defects from the observation of bound excitons, in general the most important data stem from FB and DA transitions. Figure 7.1 gives an overview of the possible processes.

Band-to-Band Transitions

In cases where the carriers in the valence and conduction bands can be regarded as moving independently, e.g., at room temperature, recombination takes place between these free carriers. Due to the motion of the free charge carriers at finite temperature, the shape of the emission intensity I_{PL} as a function of the emission energy $\hbar\omega$ is given by [9]

$$I_{\mathrm{PL}}(\hbar\omega) \propto \begin{cases} (\hbar\omega - E_G)^{1/2}\exp\left(\dfrac{-(\hbar\omega - E_G)}{k_B T}\right) & \text{for } \hbar\omega > E_G, \\ 0 & \text{otherwise} \end{cases} \qquad (7.1)$$

with the bandgap energy E_G, the temperature T and the Boltzmann constant k_B. The first term on the right-hand side reflects the root-like density of states at the band edge of a semiconductor and the second term, the Boltzmann factor of the occupation. The maximum intensity of this distribution is at an energy $E_G + k_B T/2$.

Excitonic Transitions

At lower temperatures, the Coulomb attraction between holes and electrons becomes important and excitons are formed. Their radiative recombination results in luminescence with an energy of the bandgap reduced by the binding energy of the exciton: $E_G - E_{\mathrm{FE}}$. The binding energy of the exciton E_{FE} is calculated from the hydrogen model

$$E_{\mathrm{FE}} = \frac{m_{\mathrm{r}} e^4}{2\hbar^2 (4\pi\varepsilon_0\varepsilon_{\mathrm{r}})^2} \frac{1}{n^2}, \tag{7.2}$$

where m_{r} is the reduced mass, $1/m_{\mathrm{r}} = 1/m_{\mathrm{e}} + 1/m_{\mathrm{h}}$ and m_{e} and m_{h} are the electron mass and the hole mass, respectively. The quantity e denotes the unit charge, $\hbar$ Planck's constant divided by 2π, ε_0 the permittivity of vacuum, ε_r the dielectric constant of the semiconductor at zero frequency and n is the main quantum number. Besides, in very good crystals, only the first level of the exciton with $n = 1$ is observed.

With increasing temperature, the energy position of the excitonic emission follows the temperature dependence of the bandgap. The intensity of the excitonic emission decreases as the thermal energy overcomes the excitonic binding energy. In general, the temperature dependence of emission intensity can be described by

$$I_{\mathrm{PL}} \sim \frac{1}{1 + aT^{3/2}\exp(-E_{\mathrm{act}}/k_{\mathrm{B}}T)}, \tag{7.3}$$

where a is a proportionality constant and E_{act} is the activation energy of the quenching process, e.g., in the case of an exciton, the exciton binding energy. The factor $T^{3/2}$ reflects the temperature dependence of the effective density of states at the band edge and is omitted in the case of a quenching process into a temperature-independent density of states, e.g., another defect.

The energy position of the excitonic emission does not depend on the excitation intensity I_{cx}, unless the system is dominated by band tail states. In general, the intensity of the PL emission I_{PL} follows the power law:

$$I_{\mathrm{PL}} \sim I_{\mathrm{ex}}^k \tag{7.4}$$

with the exponent $k < 2$ in the case of excitons [15]. Basically, an exponent of 2 is expected since the excitonic emission intensity depends on the concentrations p, n of holes and electrons, each of which is proportional to I_{ex}. But due to the interaction with other radiative and nonradiative recombination processes k becomes smaller than 2.

Apart from these free excitons, an exciton can form a bound state with a defect. Due to the attractive interaction between the constituents of the exciton and the defect, the binding energy of the bound exciton is increased by an amount proportional to the ionization energy of the defect, which is known as Haynes' rule [16]. The proportionality factor C depends on the ratio $m_{\mathrm{e}}/m_{\mathrm{h}}$ of the effective masses and was calculated for charged [17] and neutral defects [18]. Thus, Haynes' rule for the binding energy of the bound exciton E_{BE} reads:

$$E_{\mathrm{BE}} = \begin{cases} CE_{\mathrm{A/D}} + E_{\mathrm{FE}} & \text{for neutral defects,} \\ CE_{\mathrm{A/D}} & \text{for charged defects.} \end{cases} \tag{7.5}$$

The difference between the two is just due to the convention to express the binding energy of excitons bound to a neutral defect as the additional binding energy compared to the free exciton and for excitons bound to a charged defect as the complete binding energy of the bound exciton. Taking into account that within the hydrogen model the binding energies of, e.g., a donor and the exciton differ only by the ratio of m_r/m_e, the expression for neutral defects can be transformed into the one for charged defects.

Free-to-Bound Transitions

When there are donor or acceptor states in the bandgap a free electron in the conduction band can recombine with a hole bound to an acceptor or an electron bound to a donor can recombine with a free hole in the valence band. The shape of the luminescence reflects again the root-like density of states in the band and the thermal occupation distribution and is known as Eagles' law [19]:

$$I_{PL}(\hbar\varpi) \propto (\hbar\omega - (E_G - E_{A/D}))^{1/2} \exp\left(\frac{-(\hbar\omega - (E_G - E_{A/D}))}{k_B T}\right), \quad (7.6)$$

where $E_{A/D}$ represents the ionization energy of the acceptor E_A or the donor E_D. Besides a shift of $k_B T/2$, the maximum of the luminescence is at $E_G - E_{A/D}$. Thus, from an FB transition a determination of the defect energy $E_{A/D}$ is possible. To determine whether the defect is a donor or an acceptor, further information, for example, from Hall measurements is necessary. In certain cases, a comparison with the ionization energies obtained from the binding energies of bound excitons according to Eq. 7.5 can clarify the situation.

With increasing excitation intensity an FB transition does not change its energy position. Its intensity follows Eq. 7.4 with $k < 1$ [15], due to the fact that the PL intensity is proportional to the concentration of electrons or holes n or p, which depend linearly on the excitation intensity and PL intensity is proportional to the density of defects, which does not depend on excitation intensity. From this $k = 1$ is expected and due to the interaction with competing processes the exponent is reduced.

Donor–Acceptor-Pair Transitions

When there are donors and acceptors coexisting in the semiconductor, DA transitions will occur. Their initial state consists of a neutral donor and a neutral acceptor; their final state consists of two oppositely charged defects. Between these charged defects exists an attractive Coulomb potential. This additional energy E_{coul} is transferred to the photon, resulting in a maximum of the emission E_{PL} at

$$E_{PL} = E_G - (E_A + E_D) + E_{coul} \quad \text{with } E_{coul} = \frac{e^2}{4\pi\varepsilon_0\varepsilon_r r}, \quad (7.7)$$

where r represents the distance between the donor and the acceptor, which depends on the density of neutral donors and acceptors in the initial state, i.e., on the excitation intensity. The dependence of the energy position as a function of excitation intensity can be described by the following empirical equation [20]:

$$E_{\mathrm{PL}}(I_{\mathrm{ex}}) = E_{\mathrm{PL}}(I_0) + \beta \, \log_{10}(I_{\mathrm{ex}}/I_0). \tag{7.8}$$

The expected change in E_{coul} amounts to a few meV per decade excitation intensity. Thus, a discriminating criterion for DA transitions is a β of about 1–6 meV/decade, usually together with an exponent $k < 1$ (see Eq. 7.4). But the power dependence of the luminescence intensity is not straightforward for DA transitions: k depends on the temperature, is not necessarily constant and can also reach values up to 2 [21]. In not too complicated cases, it can be shown that the exponent of DA transitions is half that of the excitons in the same samples: $k_{\mathrm{FE}} \approx 2k_{\mathrm{DA}}$ [15].

With increasing temperature the thermal emptying of defects becomes more and more probable. For DAs with closer distance, the transition probability is higher than that for large-distance pairs due to the tunnel contribution in these transitions. Thus, the recombination of these short-distance pairs is faster and more probable than the thermal emptying [22]. Due to the Coulomb term, this leads to a small blue shift of these transitions of the order of $k_{\mathrm{B}}T$ with increasing temperature. The temperature dependence of the intensity of a DA transition follows Eq. 7.3 with an E_{act} of ionization energy of the shallower defect at lower temperatures and of the deeper defect at higher temperatures. The temperature dependence is then given by [23]

$$I_{\mathrm{PL}} \sim \frac{1}{1 + T^{3/2} \sum_i C_i \exp(-E_{\mathrm{act},i}/k_{\mathrm{B}}T)}. \tag{7.9}$$

One should keep in mind, though, that once the shallower defect is thermally emptied, the remaining FB transition between the deeper defect and the opposite band will dominate, and then of course E_{act} represents that deeper defect.

Luminescence in Highly Compensated Semiconductors

When there are comparable and high densities of donors and acceptors in a semiconductor material, most of the defects are charged, since the minority defects lose their charge carrier to the majority defect, charging both without generating free charge carriers that would otherwise screen the charges. For random spatial distributions of these defects, this situation will lead to areas with a higher density of positively charged donors and areas with a higher density of negatively charged acceptors, causing a spatially fluctuating electrostatic potential [24]. Figure 7.2 illustrates the effect.

The effects of these fluctuating potentials are described in detail in [25]. Here, only a brief summary is given. DA transitions always involve tunnel processes, since the donors and acceptors are not located at the same site.

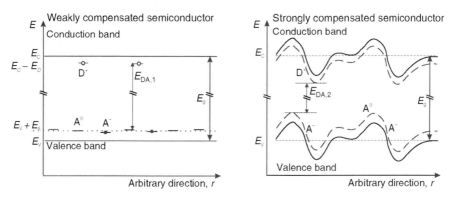

Fig. 7.2. Weakly compensated semiconductor without fluctuating potential and strongly compensated semiconductor with fluctuating potential

In a weakly compensated semiconductor, this tunnel contribution does not affect the energy of the emission (besides the Coulomb term—$E_{DA,1}$ in Fig. 7.2) because the bands are flat. In the case of bent bands due to fluctuating potentials, the tunneling between spatially separated donors and acceptors can considerably reduce the energy of the transition ($E_{DA,2}$ in Fig. 7.2). Thus, in the case of high compensation, i.e., fluctuating potentials, a broad range of energies is available for DA transitions, leading to a broadening of the energy spectrum of the emission. In addition, the average energy is lower than that in the uncompensated case, resulting in a red shift of the emission peak. Finally, there is a maximum energy that occurs for transitions with negligible tunnel distance, which is the vertical energy $E_{DA,1}$. But on the low-energy side with increasing tunnel distance, i.e., decreasing transition probability, even lower transition energies become available. This makes the emission line asymmetrically broadened, with a steeper high-energy slope determined by thermal broadening and a flatter low-energy slope determined by the fluctuation amplitude of the electrostatic potential. Thus, DA emissions under the influence of fluctuating potentials show a characteristic shape – they are asymmetrically broadened with a flatter low-energy slope and they are red-shifted compared to the undistorted emission. Another important feature of transitions under the influence of fluctuating potentials is a large blue shift with increasing excitation intensity. With increasing excitation, increasing numbers of defects are neutralized and the density of free charge carriers increases, screening the remaining charged defects. This results in a flattening of the fluctuating potentials and thus a blue shift in the energy position of the transition. Blue shifts up to 30 meV/decade of excitation intensity have been observed in GaAs and ZnSe [26–28]. This flattening of the fluctuating potentials leads to a decrease of the peak width, and finally at high excitation powers the unperturbed spectrum is obtained. With increasing temperature, the charge carriers can surmount the barriers of the fluctuating potential and different regions move into thermal equilibrium, establishing a quasi-Fermi level throughout the crystal. This results in a red shift of the emission maximum with increasing temperature.

Experimental Discrimination between Transitions

The change of energy position and luminescence intensity with excitation intensity and temperature can be used to experimentally distinguish between the different types of transitions. A summary of the above discussion is given in Table 7.1.

It should be pointed out that a major source for misinterpretation of PL spectra is the negligence of the effects of compensation and uncertainty of the nature of an emission being due to a DA or an FB transition. If a broad asymmetric transition that shows a large blue shift with increasing excitation intensity and a red shift with increasing temperature, thus showing all signatures of luminescence under the influence of compensation and fluctuating potentials, is interpreted as a normal DA transition, then the defect ionization energies concluded from its energy position are much larger than the real values. When the emission lines are broad, it can be difficult to detect small blue shifts with increasing excitation intensities. These emissions can than be interpreted as FB transitions although they are DA transitions. This leads to an overestimation of the ionization energies of the defects as well.

7.2.2 Electrical Transport Measurements by Hall Effect

From a combination of Hall voltage and conductivity measurements charge carrier density and mobility can be determined. Their temperature dependencies allow the determination of the ionization energy of the doping defects and the scattering mechanisms of the free carriers. The basic formulae on semiconductor statistics and transport can be found in numerous textbooks [9, 29–31]. Only a brief review concerning defects is given here. No detailed discussion of scattering mechanisms is given here, since we concentrate on defects.

Table 7.1. Experimental discrimination between transition types

transition type	temperature dependence		excitation intensity dependence	
	energy position	intensity (7.3)	energy position (7.8)	intensity (7.4)
excitonic	varies with E_G	$E_{\mathrm{act}} = E_{\mathrm{exc}}$	constant	$k < 2$
FB	shifts by $k_B T/2$	$E_{\mathrm{act}} = E_{\mathrm{defect}}$	constant	$k < 1$
DA (weak compensation)	small blue shift	$E_{\mathrm{act}} = E_{\mathrm{defect}}$	$\beta = $ few meV/decade	$k < 1$, usually
DA (strong compensation)	red shift		$\beta > 10\,\mathrm{meV/decade}$	

Hall Measurements

The Hall voltage U_H is used to determine the Hall coefficient R_H:

$$U_H = R_H \cdot I \cdot B / d \quad \text{with } R_H = \frac{\sum r_i q_i n_i \mu_i^2}{\left(\sum q_i n_i \mu_i\right)^2}, \tag{7.10}$$

where I is the current through the sample, B the magnetic field perpendicular to the direction of the current and d the thickness of the sample; r_i represents the Hall factor for the ith conduction path, which contains the energy dependence of scattering times characteristic for the scattering mechanism, q_i the corresponding charge of the carriers (e.g., $-e$ for electrons), n_i the respective concentration of carriers and μ_i their mobilities. The Hall coefficient is the sum over all parallel conductions paths like conduction by electrons and holes or carriers in the band and defects, r_i is usually not known, but can be calculated for a number of specific assumptions and is between 1 and 2 [29]. For a moderately doped p-type semiconductor at not too low temperatures, i.e., in the case of $p \gg n$ and conduction in the valence band as dominating path, the Hall coefficient becomes the usual $1/ep$, with the hole concentration p and assuming a Hall factor $r_i = 1$. Combining with conductivity σ measurements one obtains the mobility μ from $\sigma = \mu/R_H$.

When determining the temperature dependence of the charge carrier concentration p the fact that r_i is not known and might be also temperature dependent is usually no big problem since p changes over several orders of magnitude. But r_i is also contained in the determination of μ; and since μ changes usually not much more than one order of magnitude, the unknown r_i might be a problem here.

Charge Carrier Density

The basics of semiconductor statistics is the neutrality condition that can be written for a compensated p-type semiconductor as

$$p + N_D = \sum N_{A,i}^-, \tag{7.11}$$

where N_D is the concentration of all donors that are all ionized, even at the lowest temperatures due to compensation and $N_{A,i}$ the concentration of the ionized acceptors of type i, which is given by the Fermi distribution

$$N_{A,i}^- = N_{A,i} \left(\frac{1}{1 + g_i \exp\left(\frac{E_{a,i} - E_F}{k_B T}\right)} \right), \tag{7.12}$$

where $N_{A,i}$ is the total concentration of acceptors of type; i, g_i their degeneracy factor, which is 2 for singly charged acceptors; $E_{a,i}$ the energy position of the acceptor and E_F the energy position of the Fermi level. For the concentration of free holes the Boltzmann approximation is used

$$p = N_V \exp\left(-\frac{E_F - E_V}{k_B T}\right) \quad \text{with } N_V = 2\left(\frac{2\pi m_h k_B T}{h^2}\right)^{3/2}, \qquad (7.13)$$

where N_V is the effective density of states in the valence band and E_V the energy position of the valence band maximum. Using the ionization energy of the acceptor $E_{A,i} = E_{a,i} - E_V$ (Eq. 7.13) can be inserted into Eq. 7.12, leading to

$$N_{A,i}^- = N_{A,i}\left(\frac{1}{1 + g_i \frac{p}{N_V} \exp\left(\frac{E_{A,i}}{k_B T}\right)}\right). \qquad (7.14)$$

Inserting this into the neutrality condition Eq. 7.11 leads to a quadratic equation in p for a single type of acceptor and to a cubic equation in p for two different types of acceptors. These equations can be analytically solved, resulting in an expression for $p(T)$, which can be fitted to the experimental data, using N_D, N_{A1}, and E_{A1} (and N_{A2}, E_{A2}, where appropriate) as fit parameters. For one type of acceptor the $p(T)$ curves show the same Arrhenius-type behavior as observed in uncompensated semiconductors with a single type of defect. Only the slope is not $E_A/(2k_B T)$ as for the uncompensated case but $E_A/(k_B T)$ in the case of compensation. In the case of low compensation $K = N_D/N_A < 50\%$ both slopes are observed: $E_A/(2k_B T)$ as long as $p > N_D$ and $E_A/(k_B T)$ for $p < N_D$ (see e.g., [29]).

That the interpretation of the ionization energy E_A is not straight forward, was first discovered by Pearson and Bardeen [32] when they observed that the observed ionization energy of defects E_A in Si depends on the concentration of defects by

$$E_A = E_{A,0} - \alpha N^{1/3}, \qquad (7.15)$$

where $E_{A,0}$ defines the infinite dilution limit of the defect ionization energy, α is an empirical proportionality constant and N is the defect concentration. Various theoretical models for the description of this behavior have been put forward: screening by free charge carriers [32], influence of band tails and overlap of wave functions and band edge shifts [33] and screening by mobile charged defects [34]. Podör [35, 36] showed that Coulombic-type interactions play a major role, since $\alpha \sim \varepsilon_r^{-1}$. For a large number of different semiconductors, α has been observed to be $(4 \pm 2)10^{-8}$eV cm [35, 36]. From various theoretical approaches, it is not clear what defect concentration should be used for N. But certain experimental hints exist showing that the concentration of the minority defects is essential [37]. These two observations, Coulombic-type interaction and dependence on minority defects, could imply that in fact the screening by mobile charged minority defects plays a major role. It should be noted that the dependence of the defect ionization energy on the defect concentration has only been observed in electrical measurements, where the thermal ionization energy is determined, while it is not observed in optical investigations like PL spectra. Since according to Eq. 7.15, the measured ionization energy E_A can take any value between 0 and $E_{A,0}$ depending on the concentration of the minority defects, this represents a major source of misinterpretation. Only by investigating numerous samples with different minority defect concentrations is it possible to determine the ionization energy of the defect in the infinite dilution limit $E_{A,0}$.

Hopping and Defect Bands

At high defect concentrations, charge carriers not only move on the delocalized bands but can also move between defect states via hopping. Detailed descriptions of this transport mechanism can be found in [24, 38]. Here, only a phenomenological description of the observed effects is given.

If hopping occurs in a semiconductor, the measured charge carrier density is described in the two-path model, where the summations in Eq. 7.10 extend over the two contributions: conduction in the band (subscript b) and conduction in the defect zone (subscript d).

$$R_H = \frac{p_b \mu_b^2 + p_d \mu_d^2}{e \left(p_b \mu_b + p_d \mu_d\right)^2}, \tag{7.16}$$

where a p-type semiconductor is assumed with p_b, p_d hole concentrations in the band and in the defect zone, respectively and the Hall factor r is assumed to be 1. Using the net doping, $P = N_A - N_D = p_b + p_d$ and w, the ratio of the mobilities μ_d/μ_b, one obtains:

$$eR_H = \frac{(1 - w^2)p_b + Pw^2}{((1 - w)p_b + Pw)^2}. \tag{7.17}$$

As a function of temperature via $p_b = p_b(T)$; R_H has a maximum at $p_b = Pw/(1+w)$, i.e., when $\sigma_b = \sigma_d$. This leads to a minimum in the apparent charge carrier density. At higher temperatures, the freeze out behavior of the carriers in the band is seen; at lower temperatures, an increase in the apparent charge carrier density is observed. Basically, this is due to the increasing number of carriers in the defects, but in the hopping regime the usual Hall equations like Eq. 7.10 are not valid anymore [39–41]. Since hopping conduction involves a tunneling process between localized defect states, the measured "hopping mobilities" μ_d are usually low, of the order of or below $1\,\mathrm{cm}^2/\mathrm{V\,s}$.

In the case of very high defect concentrations, the localized defect states overlap and form a delocalized band. This insulator–metal transition occurs when the average distance between defects is of the same size as the Bohr radius a_B of the defects. A criterion for the occurrence of defect bands given by Mott [42] has been confirmed for a large number of semiconductors:

$$N^{1/3}a_B > 0.2 \quad \text{with } a_B = \frac{4\pi\varepsilon\varepsilon_0\hbar^2}{m^*e^2}, \tag{7.18}$$

where N is the density of the respective defects and m^* the effective mass of charge carriers in the corresponding band. For defect densities within this range, metallic conduction, characterized by an increase in conductivity with decreasing temperature occurs at low temperatures instead of hopping conduction. Since these defect bands are very narrow, usually a few tens of meV, the effective mass in these bands is rather high, leading to a low mobility as well.

7.3 State of the Art: Review of Defects in Chalcopyrites

This section gives an overview of the literature data concerning defects in CuGaSe$_2$ and its more intensely investigated relative CuIn(Ga)Se$_2$. Defects have been investigated by luminescence spectroscopy and by Hall measurements. The deeper defects found in capacitance measurements on devices are not discussed here. This section concentrates on the defects responsible for doping.

As discussed in the previous sections, luminescence spectra, whether cathodoluminescence or PL, and temperature-dependent Hall measurements provide information about the donors and acceptors and their ionization energies in a semiconductor. A critical summary of previous investigations is given.

7.3.1 Defects in CuInSe$_2$

The first PL spectra of CuInSe$_2$ crystals were measured in the early 1970s at Bell Labs [43], where the first chalcopyrite solar cells were also prepared from CuInSe$_2$ single crystals [44]. This first PL study was used to investigate the band structure and not to identify defects. An earlier PL study concerned with defects and dating back to 1975, identified an acceptor 40 meV deep in p-type (Se-rich) and n-type (Se-poor) material and a donor 70 meV deep in n-type material [45]. In 1972, the first Hall measurements indicated that Se-poor CuInSe$_2$ shows p-type behavior and Se-rich material n-type behavior [46]. The first native defect found by Hall measurements on single crystals was a shallow donor around 10 meV [47]. A literature summary of the defects found since then in CuInSe$_2$ by luminescence investigation and Hall measurements is given in Fig. 7.3. The last values are from our group and will be discussed in detail later.

There exist many more publications than are included in the figure, but not all of them are reliable. The first problem arising in luminescence measurements is the difficulty to distinguish between acceptors and donors, although we will see that with careful analysis and using Hall data an attribution can be made. Another difficulty is the neglect of the effects of fluctuating potentials. Even if the transitions are clearly asymmetric and show a high blue shift with

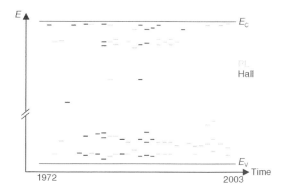

Fig. 7.3. Defects found in CuInSe₂ by luminescence and Hall measurements taken from [45,47–77]

excitation intensity, often the transitions are interpreted as normal DA transitions, resulting in very deep defects. Another problem occurring regularly is the neglect of the Coulomb term in DA transitions. For the determination of defect ionization energies information on the bandgap is essential. It cannot be taken from the literature, especially in the case of thin films, since the bandgap varies with strain in the film. Therefore, it is necessary to detect the excitonic emissions and make correct assumptions on the exciton energy. As seen from Eq. 7.2, knowledge about the effective mass and the dielectric constant of the material is required. For a discussion of literature values see the appendix.

The main problem with Hall data is the concentration dependence of the defect ionization energy. The only value that can be used for comparison with other data is the infinite dilution limit, which can only be determined from a large number of samples. Therefore, most of the Hall data that do not take into account this screening effect will indicate very low values of the defect ionization energies.

Even more problematic than the pure defect energies are the defect chemical correlations. They are mostly based on compositional trends, on temper experiments and on theoretical calculations of defect formation energies (see [54] based on data by Neumann) and defect energies [78]. Specifically, temper experiments that do not take into account the ternary character of the material cannot be used for defect chemical identification [72, 79]. And most temper experiments treat these ternary materials as if they were binaries assuming one of the components does not change.

Summarizing the luminescence and Hall data in Fig. 7.3, one might conclude that there exist two to three donor defects in CuInSe₂ – a shallow one below 10 meV and one or two in the region between 60 and 90 meV from the conduction band, two different acceptor states – one at 30–40 meV and a deeper one at 70–100 meV; and maybe a third one around 150 meV.

However, one must keep in mind that this accordance is at least in part due to self-reference. Another point is that the interpretation of the luminescence spectra leads either to the conclusion of donors in the 60–90 meV region or of shallow donors, where the assumption of shallow donors is usually associated with the occurrence of the deeper acceptor.

Nevertheless, certain analogies can be found in the reported PL spectra, at least in those cases where narrow, symmetric emission are observed: Excitonic emission is usually found around 1.03–1.04 ev (see e.g., [48, 52, 58, 80]). A first defect-related emission around 0.98–1.00 ev, which is attributed either to an FB transition or a DA transition (see e.g., [45, 48, 50, 52, 56]) and a second DA/FB emission around 0.97 ev (see e.g., [45, 48, 50, 52, 58, 63]). In PL measurements these two transitions either occur together or singly. Until recently, it was not clear what experimental conditions make them appear or disappear. Usually, they are accompanied by transitions at lower energies at a distance of 29 meV, which are attributed to phonon replica. A number of studies find a deeper DA/FB transition around 0.89–0.92 eV [45, 52, 56, 58, 63]. Although the spectra are similar, the interpretations vary widely. Thus, it must be stated that after more than 30 years of research on the defects in CuInSe$_2$ no consistent picture has emerged.

7.3.2 Defects in CuGaSe$_2$

Much less luminescence and Hall investigations have been done on the wider bandgap compound CuGaSe$_2$. A literature survey is given in Fig. 7.4. The last three are data from our group and will be discussed in detail later.

The main difference between Hall measurements of CuInSe$_2$ and CuGaSe$_2$ is that measurements of CuGaSe$_2$ cover only acceptors, due to the fact that

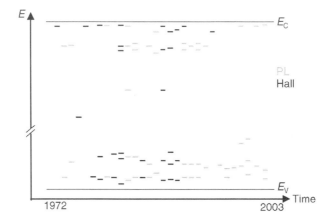

Fig. 7.4. Defects found in CuGaSe$_2$ by luminescence and Hall measurements taken from [57, 81–95]

unlike CuInSe$_2$ under all deviations from stoichiometry CuGaSe$_2$ is always p-type [46].

Summarizing the results from Fig. 7.4, it appears that there might be two to three donors again: one or two in the energy range between 30 and 80 meV and a deeper one at 110 meV, and probably two acceptors at 30–60 and 110 meV. It should be noted that the 110 meV defect appears alternatively as an acceptor or a donor. In the case where it appears as an acceptor and a donor [83], it was discussed as either one of the two. The same problems of self-reference and interpretation of the data apply as are discussed for CuInSe$_2$. The same conclusion must be drawn: after 25 years of research no consensus has been reached.

Nevertheless, there are certain agreements considering the observed PL transitions, disregarding the varying interpretations – an emission around 1.71–1.73 ev is usually interpreted as excitonic emission (see e.g., [80, 82, 84, 85, 87, 90, 96–99]). This is followed by either of two or both transitions around 1.66–1.69 and 1.62–1.63 ev, which are interpreted as FB or DA transitions (see e.g., [80, 82–87, 90, 91, 97, 99–105]), although some authors report more individual transitions in this energy region. A deeper defect-related transition around 1.56–1.60 ev has been reported as well (see e.g., [81, 83–85, 91, 100, 101, 103–105]), together with a number of deeper transitions, which have been attributed to deeper defects [100, 106] or to potential fluctuations [103]. These general patterns are similar to those obtained in CuInSe$_2$ – a fact that has been ignored in the literature until recently [77, 107].

Comparison of CuGaSe$_2$ and CuInSe$_2$

The differences between CuGaSe$_2$ and CuInSe$_2$ discussed in Chap. 1, like CuGaSe$_2$ being always p-type under any stoichiometry deviation or the much lower efficiency of solar cells with CuGaSe$_2$ absorber, could imply that the defect spectrum of these two materials is different. Only very few authors attempt a direct comparison between the defects in CuInSe$_2$ and CuGaSe$_2$; examples are given in Table 7.2. It appears as if there are less defects detected in CuGaSe$_2$ than in CuInSe$_2$, but it is unclear whether this is due to the fact that CuGaSe$_2$ is less investigated or whether this is a fundamental difference.

Table 7.2. Literature data comparing defects in CuInSe$_2$ and CuGaSe$_2$

defect type	energies in CuInSe$_2$ (meV)	energies in CuGaSe$_2$ (meV)	reference
donor	10–20, 70, 200	80, 110	[108]
acceptor	30, 40, 90	30–50, 150	
donor	60, 80	80, 110	[55]
acceptor	40, 80	50	

7.3.3 Defects in Cu(In,Ga)Se$_2$

Only very few papers are available on defect spectroscopy in the alloy. A first luminescence study by Massé showed, at least within their recombination model, that defect ionization energies can be interpolated between CuInSe$_2$ and CuGaSe$_2$ [57], but based on rather broad luminescence peaks. The work by Wagner and Dirnstorfer and coworkers [63, 64, 109–111] on Cu(In,Ga)Se$_2$ prepared by a sequential process draws conclusions on defects mostly for CuInSe$_2$ since the spectra they observe consist of CuInSe$_2$ spectra plus a broad background from Cu(In,Ga)Se$_2$. An important conclusion of their work is that the broad asymmetric luminescence peaks in pure CuInSe$_2$ due to high compensation and fluctuating potentials change to the narrow lines of material with low compensation after annealing in air, while they remain broad for Cu(In,Ga)Se$_2$ with a [Ga]/[Ga] + [In] ratio above 10%. These fluctuating potentials were also interpreted as lateral composition fluctuations [112]. The broad luminescence was observed in polycrystalline films as well as in single crystals [113, 114]. A recent time-dependent study showed for Cu(In,Ga)Se$_2$ polycrystalline films with a [Ga]/[Ga] + [In] ratio of about 30% that the lifetime of luminescence is much longer for Cu-poor films than for Cu-rich films [115]. A Hall study on epitaxial Cu(In,Ga)Se$_2$ indicates two acceptor levels around 170 and 40 meV for low Ga content, the defect ionization energies decreasing with increasing Ga content [116]; the influence of hopping conduction was also discussed in the study.

7.4 Defect Spectroscopy in CuGaSe$_2$

This section describes detailed and systematic defect spectroscopic measurements by PL and Hall investigations on a large number of CuGaSe$_2$ and CuInSe$_2$ samples with varying composition allowing probative conclusions on the defects in chalcopyrites.

7.4.1 Sample Preparation

Epitaxial and polycrystalline films of CuGaSe$_2$ and epitaxial films of CuInSe$_2$ are deposited by metal organic vapor phase epitaxy (MOVPE) and by physical vapor deposition (PVD), i.e., thermal evaporation.

MOVPE is performed on GaAs(0 0 1) substrates in a horizontal quartz reactor. The precursors are cyclopentadienyl Cu triethylphosphine or cyclopentadienyl Cu tertiarybutyl isocyanide, triethyl Ga, trimethyl In and ditertiarybutyl Se using Pd-diffused hydrogen as a carrier gas. Due to the low vapor pressure of the Cu precursors, the Cu source is operated at an elevated temperature of 60°C and passed into the reactor separately from the other metal organics in a heated line. The reactor pressure is 50 mbar and the substrate temperature is 570°C for CuGaSe$_2$ and 500°C for CuInSe$_2$. These recipes basically follow the procedure developed by Chichibu [117]; a detailed discussion can be found in [118,119]. The growth rates are low, around 0.3 Å/s, and the thickness of the epilayers is typically 0.5 μm.

Aiming at a controlled variation of the chemical composition of the chalcopyrite epilayers, the input partial pressure ratio Cu-precursor/(Ga-precursor or In-precursor) (p_I/p_{III} ratio) is varied, while the Se-precursor flow ratio ($p_{VI}/(p_I + p_{III})$) is kept constant at ~10. These growth experiments are performed using substrate rotation to ensure homogeneity of the composition across the samples.

The resulting chemical composition of the epilayers is determined by energy dispersive X-ray spectroscopy (EDX) and X-ray fluorescence (XRF). The [Cu]/[III] ratio given in the following is measured for as-grown epilayers integrating over the ternary chalcopyrite phase and any secondary phases that might segregate according to the phase diagram [1,120]. Samples characterized by [Cu]/[III] ≥ 1 consist of a near-stoichiometric ternary chalcopyrite phase and Cu$_x$Se segregating due to Cu-excess.

For CuGaSe$_2$ and CuInSe$_2$ epilayers grown at different p_I/p_{III} ratios, a linear dependence between the mole fraction [Cu]/[III] in the solid and the partial pressure ratio p_I/p_{III} in the vapor was found, thus confirming good controllability of the composition in the MOVPE processes.

Polycrystalline and epitaxial films of CuGaSe$_2$ are also grown by PVD onto soda lime glass and GaAs(0 0 1), respectively. The PVD system is equipped with two thermally stable effusion cells for Cu and Ga. Se is evaporated from a tungsten boat or from an effusion-cell-type evaporator. The Se flux is regulated by a quartz crystal monitor controlling the power applied to the source. The base pressure in the evaporation chamber is below 10^{-6} mbar and during the process below 5×10^{-6} mbar. A 10 kW halogen lamp heater system with low thermal mass allows fast heating and cooling ramps of the substrates. For homogeneity the substrate holder is rotated. Cu-poor films are grown by a bi-layer process, similar to the one originally used by Boeing [121] with a slightly Cu-rich growth in the first stage and deposition of only Ga and Se in the second stage to get an overall Ga-rich composition. Cu-rich films are grown in a single-stage coevaporation process with constant evaporation rates throughout the process time. Se is supplied in high excess until the substrate has cooled down to 250°C. The substrate temperature is ramped from 350°C up to the softening point of the glass within the first 15 min of deposition. Growth rates are in the range of 7 Å/s and the typical thickness of polycrystalline layers is 1.6 μm; for epitaxial layers the thickness is 0.5 μm. Solar cells made from Ga-rich polycrystalline absorbers grown by the same process reach energy conversion efficiencies of up to 8.2% [122], not much lower than the world record for thin-film solar cells with a CuGaSe$_2$ absorber of 9.5% [8].

7.4.2 Photoluminescence Spectra of CuGaSe$_2$

In this section, the PL spectra of epitaxial CuGaSe$_2$ layers will be discussed. It should be noted that the same type of spectra and the same composition dependencies are observed in polycrystalline layers.

For the PL measurements, the 514.5 nm line of a cw Ar$^+$ laser is used as an excitation source. Neutral density filters are used to provide a range of excitation intensities between 1 and 100 W cm^{-2}. Samples are cooled down to 10 K in a helium bath cryostat. Emission spectra are analyzed by a double-grating monochromator with a focal length of 0.22 m and detected phase sensitively using a photomultiplier or an InGaAs diode, depending on the spectral range.

Figure 7.5 shows PL spectra of CuGaSe$_2$ layers measured at 10 K as a function of composition [92]. The integrated [Cu]/[Ga] ratios of the epilayers are given in Fig. 7.5 on the left-hand side ranging from 0.6 < [Cu]/[Ga] < 1.2. The observed PL properties as a function of [Cu]/[Ga] ratio can be summarized as follows:

1. *[Cu]/[Ga] > 1*: In the case of films grown under Cu excess, excitonic luminescence is observed [123]. Two optical transitions involving defects dominate the spectra showing line widths of about 20 meV. Their peak energies of 1.62 and 1.66 ev appear to be independent of Cu excess, but their intensity ratio changes systematically with increasing [Cu]/[Ga] ra-

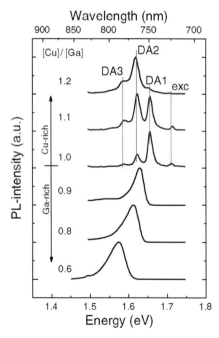

Fig. 7.5. Photoluminescence spectra of CuGaSe$_2$ with different compositions

tio. These emissions are due to DA transitions, as is discussed in the following. The third defect-related peak was previously identified as a phonon replica [92]. But, as the following discussion shows, it is in fact a third defect-related transition.

2. $[Cu]/[Ga] < 1$: The PL spectra of Cu-poor CuGaSe$_2$ show single, broad emission bands with an asymmetric line shape characterized by a high-energy slope that is steeper than the low-energy slope. With increasing Ga content these emission bands broaden and their peak maxima shift to lower energies.

These dependencies were confirmed in numerous PL measurements with epitaxial layers grown under varying conditions and with polycrystalline layers.

Characteristics of PL Emissions of CuGaSe$_2$, Grown Under Cu-Excess

CuGaSe$_2$ grown under Cu-excess consists essentially of near-stoichiometric chalcopyrite plus Cu selenide in a secondary phase accommodating the Cu-excess. The Cu selenide phase can be easily etched away. Besides, a small energy shift due to strain release the PL spectra do not change after the etch. Thus, they are due to the nearly stoichiometric chalcopyrite.

Excitonic Emissions

Since bound excitons provide information on defect ionization energies it is worthwhile to have a look at excitonic emissions. Figure 7.6 shows the excitonic emissions at low temperature. In these single crystals, it was even possible to detect the first excited state of the exciton ($n = 2$). Thus, from Eq. 7.2 it is possible to determine the exciton binding energy from the energy difference between the $n = 1$ and the $n = 2$ emission to 13 ± 2 meV [123]. The bound exciton has an additional binding energy of 4.4 meV. Table 7.3 summarizes the constant C in Eq. 7.5, according to [17, 18], for a ratio of the effective masses m_e/m_h of approximately 0.12 (see appendix) and the corresponding defect ionization energies obtained from Eq. 7.5. Thus, in all conceivable cases, the ionization energy of the defect to which the exciton is bound to, is around 10 meV; besides, in the case where the exciton is bound to a neutral acceptor, the ionization energy amounts to ~ 60 meV. Figure 7.6(top) shows the defect-related emissions from the crystal as well. The spectrum is dominated by an FB transition. The ionization energy of the defect involved in this transition can be determined from Eq. 7.6 and amounts to 66 ± 10 meV. No emissions that would correspond to a defect with an ionization energy of around 10 meV in this crystal are observed. Therefore, it is concluded that the exciton is bound to an acceptor with an ionization energy of 60 meV.

Table 7.3. Results of Hayne's rule for CuGaSe$_2$

defect	A^0	D^0	A$^-$	D$^+$
C	0.07	0.29	1.50	1.05
ionization energy (meV)	63	15	11	16

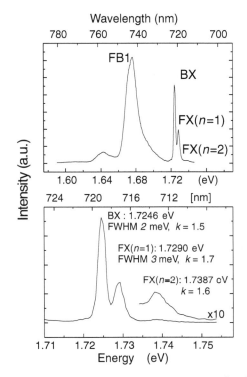

Fig. 7.6. Excitonic emission from a CuGaSe$_2$ single crystal [123]. The BX and FX ($n = 1$) emissions are observed identically in epitaxial films. The peak positions as well as the line width and the power exponents are given according to Eq. 7.4

Defect-Related Emissions

The two dominant emission lines in CuGaSe$_2$ grown under Cu-excess are attributed to DA transitions based on excitation intensity-dependent PL experiments. Figure 7.7 shows PL intensity and peak energies for both transitions with the excitation intensity varying over three decades.

The intensity I_{PL} follows the power law of Eq. 7.4 with an exponent $k < 1$ that is typical for a defect-correlated recombination mechanism (see Table 7.1). A blue shift β of the energy position with increasing excitation intensity in the range of 1–2 meV/decade indicating a DA recombination is observed.

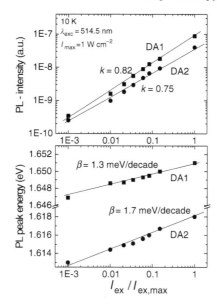

Fig. 7.7. Excitation intensity, dependent intensity and energy position of the two dominant DA transitions [92]

From the energy position of the DA transition alone it is not possible to determine defect ionization energies. As seen in Eq. 7.7, the peak position depends not only on the sum of donor and acceptor ionization energies, but also on the Coulomb term, which in turn, depends on the excitation power density. Therefore, temperature-dependent measurements were performed to obtain defect activation energies. Figure 7.8(a) shows the temperature dependence of the normalized peak, evolving from the DA1 emission.

At 10 K, the DA1 transition dominates the spectrum. With increasing temperature the line shape of the emission band changes characteristically: the DA1 transition is attenuated gradually, whereas a second transition at about 1.67 ev emerges on the higher-energy side. The origin of this latter transition was analyzed by comparing the shape of the high-energy slope with the FB recombination model by Eagles Eq. 7.6. Figure 7.8(b) shows a comparison between the experimentally observed luminescence line shape (open circles) and the profiles calculated from Eq. 7.6 for three different temperatures (lines). Excellent agreement between experiment and theory is obtained for the peak positions and the high-energy edge of the spectral line. The deviation on the low-energy tail reflects the presence of the DA1 transition and the fact that the simple theory neglects band tailing. Within uncertainty, the temperatures T_e determined from the calculation agree with the temperatures of the samples measured in the cryostat reflecting the fact that the electrons are in thermal equilibrium with the lattice before they recombine. The line shape analysis, therefore, confirms the contribution of free carriers

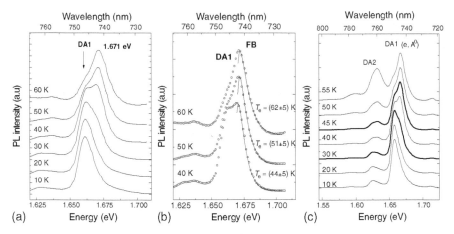

Fig. 7.8. Temperature dependence of the DA peaks in CuGaSe₂

to the recombination mechanism, which is consequently assigned to an FB
transition – the same as was observed in the single crystal in Fig. 7.6.

The temperature-dependent change of the DA1 line shape is thus interpreted as a change of the dominant recombination mechanism from DA
recombination at lower temperatures to FB recombination at higher temperatures. This process is driven by the thermal ionization of the shallower level
involved in the DA recombination, thus enhancing the probability for the
recombination of free carriers with the remaining deeper defect level. From
the calculated line shape of the FB transition, the ionization energy of the
defect level involved can be determined to be $E_{A/D} = 60 \pm 10$ meV (using the
low-temperature E_G of 1.73 ev [123]). The ionization energy of the shallower
defect involved in the DA1 transition can be estimated from the energy difference between the DA peak and the FB peak (9 meV) and an estimate of
the Coulomb energy. Defect densities obtained from Hall measurements (see
Sect. 7.4.4) are around 10^{17} cm^{-3}, of those are less than 10^{16} cm^{-3} excited by
the intensities used here. This results in an average distance between excited
donors and acceptors of $(\pi/2N_D)^{-1/3} > 40$ nm [14], resulting in a Coulomb
energy of <3 meV (7.7). Thus, the shallow defect can be estimated to be
12 meV deep.

This leaves the question of the assignment of donor and acceptor charac-
ters. We contribute the 60 meV defect to an acceptor and the 12 meV defect
to a donor on the basis of the following arguments:

- The hydrogen model of the defect suggests donors with lower ionization
  energies than those of acceptors due to the lower effective mass of electrons
  compared to holes (compare to Eq. 7.2).
- The energy of the bound exciton observed in our films suggests as partner
  an acceptor of about 60 meV ionization energy (see the previous section).

- In high-quality single crystals, only the FB transition as observed in Fig. 7.8 is obtained, not the DA transition (see Fig. 7.6); these crystals are p-type, thus the defect involved in the FB transition must be an acceptor.

Exactly the same temperature-dependent behavior is obtained for the DA2 peak [92]. At low temperatures, it is dominated by the DA behavior (Fig. 7.8c). At 30 K, where the FB1 peak appears as a shoulder of the DA1 peak, also a shoulder appears at the high-energy side of the DA2 peak. It evolves into a clear FB transition with approximately the same energy difference between DA and FB. The fact that the transition from DA character to FB character appears at exactly the same temperature for both luminescence transitions shows that the same shallow defect that is thermally emptied is involved in both the transitions. From the energy position of the FB transition observed at higher temperatures the ionization energy of the remaining defect is determined to be 100 meV. Since the shallow defect has been attributed to a donor, this 100 ± 10 meV defect is an acceptor as well, in contrast to previous attributions to a donor-type defect [83,85]; this has been in some cases revised by the authors themselves and has been attributed to an acceptor [86,124].

The third defect-related peak, labeled DA3 in Fig. 7.5, was previously attributed to a phonon replica of DA2 [92]. Recent time and spatially resolved cathodoluminescence investigations have shown that DA3 is not correlated to DA2 or DA1, but is a transition involving a separate defect [95]. The intensity-dependent behavior of the peak energy and intensity of DA3 are shown in Fig. 7.9. The small blue shift of 1.4 meV/decade excitation intensity and the power exponent of 0.87 show that DA3 is not only a defect-related transition, but also a donor–acceptor transition as well. Figure 7.10 shows the temperature dependence of the DA3 transition. As observed for DA1 and DA2, with increasing temperature a second peak appears at the high-energy side. The separation is not as good as for DA1 and DA2, due to the broader line shape of DA3 and since DA3 appears as a shoulder of DA2. The transition between the DA and the FB occurs again between 30 and 40 K, indicating that it is again the same shallow donor that is thermally emptied, thus causing the

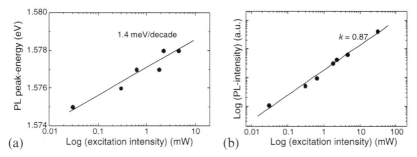

Fig. 7.9. Intensity dependence of the energy position (a) and intensity (b) of DA3 together with fits to Eqs. 7.8 and 7.4

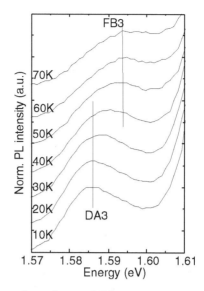

Fig. 7.10. Temperature dependence of DA3; *vertical lines* mark the peak position of the DA and the corresponding FB transition

transition from DA character to FB character. From a two-Gauss fit to the spectrum of FB2 and FB3 the peak position of FB3 can be determined as 1.593 eV. With the low-temperature gap energy $E_G = 1.73$ eV, this results in an ionization energy of the remaining acceptor of 135 ± 10 meV.

In conclusion, the spectrum of the shallow defects as obtained from PL spectroscopy consists of a shallow donor about 12 meV deep and three acceptors 60 ± 10, 100 ± 10 and 135 ± 10 meV deep. The two shallower ones dominate the spectrum, with the 100 meV one dominating for high Cu-excess and the 60 meV one dominating for low Cu-excess. The recombination model for Cu-rich $CuGaSe_2$ is summarized in Fig. 7.11. The presence of the 135 meV defect appears mostly independent of the Cu-excess.

Characteristics of PL Emissions of Cu-Poor $CuGaSe_2$

$CuGaSe_2$ films grown under Cu deficiency represent actual Cu-deficient chalcopyrite material with much more defects to be expected than in the near-stoichiometric material grown under Cu-excess. The higher number of defects is reflected in the higher intensity of radiative recombination between defects as seen in Fig. 7.12, where the absolute intensities (not normalized as in Fig. 7.5) are shown on a logarithmic scale. Both were measured under comparable conditions; the difference in intensity, which might be caused by different excitation and optical conditions is no more than a factor of 2. The intensity of the Cu-poor film is about two orders of magnitude higher than that of the film grown under Cu-excess.

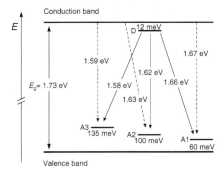

Fig. 7.11. Recombination model for CuGaSe₂ based on PL measurements; *solid lines* represent DA transitions and *dashed lines* FB transitions

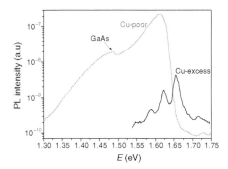

Fig. 7.12. Comparison of absolute intensities of Cu-poor CuGaSe₂ and CuGaSe₂ grown under Cu-excess

Films grown under Cu-poor conditions show a completely different PL behavior than those grown under Cu-excess: a single asymmetrically broadened emission line that shifts towards lower energies with increasing Cu deficiency (Fig. 7.5). The shape is typical for DA transitions under the influence of fluctuating potentials, as is the large blue shift with increasing excitation intensity and the red shift with increasing temperature as shown in Fig. 7.13 (cf. Table 7.1). Several clear trends are observed with increasing Cu deficiency: the peak energy shifts red, the peak width broadens and the blue shift with intensity (β) increases. All of these are indicative of an increasing influence of fluctuating potentials (cf. Fig. 7.2) with their amplitude increasing with increasing Cu deficiency, i.e., increasing defect density. The expected trend of the red shift with temperature increasing with increasing Cu deficiency is not very clear, but this might be due to slight differences of the excitation intensity, which have a large influence on the fluctuating potential amplitude, as discussed in Sect. 7.2.1.

Thus, in conclusion, the PL emission of Cu-poor CuGaSe₂ are red shifted not due to the appearance of new deeper defects but due to the existence of

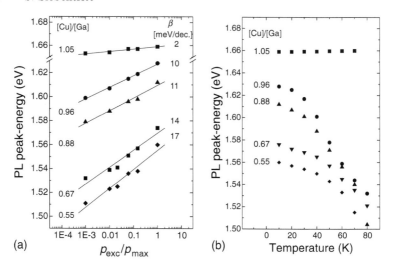

Fig. 7.13. Excitation intensity and temperature dependence of the emission peak energy of Cu-poor CuGaSe$_2$ [92]

fluctuating potentials that are caused by the high defect density and the high degree of compensation in this material.

Comparison with Literature Data

The recombination model put forward in Fig. 7.11, and particularly its compositional dependence can be used to interpret most of the literature data discussed in Sect. 7.3.2. As most authors do, we interpret the emission around 1.71 ev as excitonic transitions. Depending on the exact composition of the investigated sample, the spectrum is dominated by DA1 or DA2. Thus, there is no contradiction between reports on PL spectra dominated by a transition around 1.62 ev [83, 85, 113] and those on spectra dominated by a transition around 1.67 ev [84, 90, 97, 102]. They were just measured on samples grown with different Cu-excesses. The different interpretations as DA or FB transitions can be partly attributed to the measurement temperature and can in some cases be due to the fact that the small blue shifts of a few meV are not observable if the peaks are rather broad. Some authors have also reported on deeper defects with emissions around 1.59 ev, which can now be interpreted as the DA3 transition. Additional deeper emissions are most likely due to even deeper defects, which have not been observed in our films.

7.4.3 Comparison with CuInSe$_2$

Having analyzed the defect spectrum of CuGaSe$_2$, the question arises whether there are fundamental differences between the defect spectra of CuInSe$_2$

and CuGaSe$_2$, which could explain the observed differences discussed in Sect. 7.1. Therefore, the same detailed composition, temperature and intensity-dependent study of the PL spectra of CuInSe$_2$ are performed [77, 107] on material that is grown in the same MOVPE apparatus under similar conditions, to exclude differences due to different growth conditions.

Figure 7.14 gives an overview of the PL spectra of CuInSe$_2$ grown under basically the same conditions as the CuGaSe$_2$ samples and measured under the same conditions as the spectra shown in Fig. 7.5. The composition range covered here is somewhat smaller than that covered for CuGaSe$_2$. The reason is that luminescence from higher gap ordered defect compounds is detected in CuInSe$_2$ at [Cu]/[In] ratios lower than approximately 0.85. Comparing Figs. 7.14 and 7.5, it becomes clear that the PL spectra of the two materials are very similar: narrow emission lines for the material grown under Cu-excess and red-shifted and asymmetrically broadened emissions in Cu-poor CuInSe$_2$. This analogy will be confirmed by a closer analysis as discussed in the following. One difference occurs immediately, though: the narrow emission typical for the nearly stoichiometric material grown under Cu-excess occur down to a [Cu]/[In] ratio of 0.94. This correlates with the phase diagram of CuInSe$_2$, where Cu selenides start to form already below the stoichiometric ratio [125].

The emissions of CuInSe$_2$ grown under Cu-excess can be used to determine defect ionization energies. To analyze the character of the observed transitions, excitation-dependent PL spectra have been measured. The peak labeled "exc" has been attributed to excitonic emission. It consists of at least two peaks, which can be separated by intensity and temperature-dependent

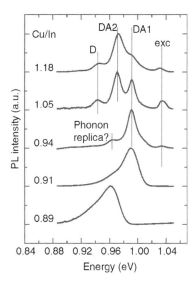

Fig. 7.14. Photoluminescence spectra of CuInSe$_2$ with different compositions

measurements. Other authors have found much better resolved excitons, they found three bound excitons in addition to the free one in the same energy region [80].

From the exponent $k \approx 1.1$ obtained according to Eq. 7.4, it is concluded that here too, emission due to excitons is obtained. From the exciton emission peak energy, the bandgap can be estimated according to Eq. 7.2 to 1.045 ev at the measurement temperature of 10 K.

The excitation intensity dependence of emission intensity and peak position of the two dominating peaks labeled "DA" is shown in Fig. 7.15 . Since the peaks are closer together than in CuGaSe$_2$ and it cannot be excluded that the intensity dependencies of the two peaks influence each other, the exponent k and intensity blue shift β are determined from different spectra: the values for DA1 from a sample with low Cu-excess, dominated by DA1, and the values for DA2 from a sample grown under high Cu-excess. Both show a slight blue shift with excitation intensity and an exponent below 1. Thus, it can be concluded that they represent DA transitions.

A third defect-related peak labeled D has been previously interpreted as a phonon replica. Based on the results obtained for CuGaSe$_2$, it is reasonable to assume that this is also a third DA transition. Its intensity dependence yields a blue shift $\beta = 1.6 \pm 2$ meV/decade and a power exponent $k = 0.39$, both compatible with the assumption of a DA transition. A determination of the energies of the involved defects turns out to be very difficult, since the peak disappears quickly with increasing temperature and neither the peak energy

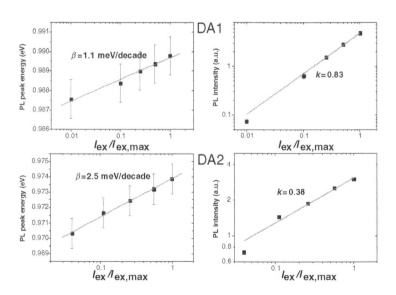

Fig. 7.15. Excitation intensity dependent intensity and energy position of the two DA transitions in CuInSe$_2$ [107]

nor the intensity can be determined reliably. Further dedicated measurements
are needed.

The fourth peak appearing in the near-stoichiometric spectrum, labelled
"phonon replica?" has an energy distance to the DA1 peak of 26 meV, which
corresponds to the energy of the LO phonon of 29 meV [126]. Therefore, it
can be assumed that this peak is in fact a phonon replica.

To determine defect ionization energies of the defects involved in DA1
and DA2, temperature-dependent measurements have been performed. A blue
shift of the DA peaks together with some broadening is observed (Fig. 7.16),
indicative of the transition to an FB emission. But no distinct FB peak is
observed; therefore, it is not possible to determine whether the shift from the
DA to the FB is completed and thus it is not possible to determine the defect
ionization energy from the peak position. One thing can be observed from the
temperature-dependent spectra: the blue shift starts at the same temperature
for both DA transitions, at 25 K. Therefore, it is concluded that in both cases
it is the same shallow defect that starts to be thermally emptied. Defect ionization energies can be determined from quenching curves according to Eq. 7.9.
They are shown together with a fit to Eq. 7.9 and the resulting activation
energies in Fig. 7.17. The deeper defect involved in DA1 shows an activation
energy of 42 meV. From an estimate of the Coulomb energy and the energy
difference between the DA peak and E_G, the energy of the shallow defect
can be estimated to be ≥ 12 meV. It has not been possible, with the current
samples, to measure the intensity of the DA2 emission over a larger intensity
range; therefore, errors in the activation energies of DA2 are much larger.

The quenching curve of DA2 cannot be described by the emptying of a
single defect. This is due to the larger ionization energy of the deeper of the
two defects involved in DA2 (therefore the lower emission energy), which leads
to an onset of the emptying at higher temperature and therefore the influence
of the shallower defect becomes visible. The deeper defect has an ionization
energy of approximately 60 meV. The lower activation energy observed in the

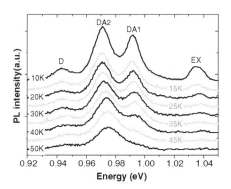

Fig. 7.16. Temperature-dependent PL spectra of CuInSe₂, grown under moderate
Cu-excess [107]

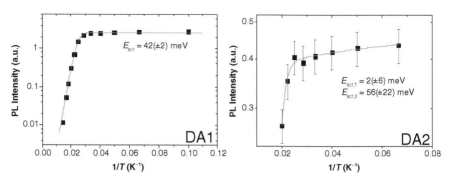

Fig. 7.17. Quenching curves of the DA transitions in CuInSe$_2$ together with the fits according to Eq. 7.9 [107]

DA2 quenching curve has a huge error. Therefore, we assume it is due to the same shallow defect as in DA1: the 12 meV donor.

According to the hydrogen model for defects in semiconductors, taking into account the much lower effective mass of the electrons, the shallower defects should be the donors, while the deeper ones are expected to be acceptors. Therefore, we conclude that the defects close to the band edges in CuInSe$_2$, as observed by PL, are two acceptors 40 and 60 meV deep with the shallower one dominating for low Cu-excess and the deeper one for high Cu-excess, and a donor approximately 12 meV deep. From the transition observed around 0.95 ev, one can assume the existence of a third deeper acceptor (see Fig. 7.18). Thus, it is concluded, comparing Figs. 7.11 and 7.18, that the shallow defects in CuInSe$_2$ and CuGaSe$_2$ are the same. The defects in CuInSe$_2$ are somewhat shallower than those in CuGaSe$_2$ as is predicted by the hydrogen model. This recombination model can again be used to interpret the PL spectra in the literature, taking the compositional dependence into account [107].

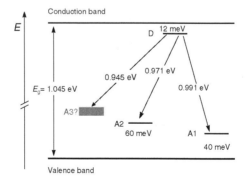

Fig. 7.18. Recombination model for CuInSe$_2$ based on PL measurements

In Cu-poor CuInSe$_2$, as in Cu-poor CuGaSe$_2$, red shifted and asymmetrically broadened emissions are observed (Fig. 7.14). As in CuGaSe$_2$ the intensity of the emissions from Cu-poor material is about an order of magnitude higher than the intensity of emission from CuInSe$_2$, grown under Cu-excess. Intensity-dependent measurements show blue shifts β of more than 12 meV/decade, typical for DA transitions under the influence of fluctuating potentials. All these observations together lead to the conclusion that in the Cu-poor regime in CuInSe$_2$, as in CuGaSe$_2$, no new (deeper) defects occur, but the emissions are red shifted due to fluctuating potentials.

7.4.4 Defect Spectroscopy of CuGaSe$_2$ by Hall Measurements

The shallow defects observed in PL investigations are the ones that are responsible for the doping of material. Therefore, they should be observable in Hall measurements as well. Hall and conductivity measurements are performed on CuGaSe$_2$ grown under Cu-excess [93]. All CuGaSe$_2$ samples are p-type, as expected.

Defect Ionization Energies

Here, we analyze the temperature-dependent charge carrier concentrations of a large number of samples, polycrystalline and epitaxial, grown by different methods (MOVPE and PVD). A selection of measured $p(T)$ dependencies is shown in Fig. 7.19. Besides the substrate and the preparation method the samples differ in the Cu-excess during growth.

Three different types of curves can be distinguished: (1) those following an exponential freeze out behavior down to the lowest measurement temperatures; (2) those with a small deviation from the freeze out behavior with somewhat higher apparent carrier concentration but still monotonically falling with decreasing temperature; and (3) those with a minimum in the apparent carrier concentration followed by increased concentration toward lower temperatures.

The latter is exactly the behavior that is expected for a transport mechanism with two paths, band transport and defect (hopping) transport, having different temperature dependencies. In such a two-path system the Hall coefficient R_H is given by Eq. 7.16 with a minimum in the apparent charge carrier concentration. Samples with type (3) temperature-dependent carrier concentration show the minimum, while type (2) samples only start to show the influence of hopping conduction; the minimum would occur at lower temperatures. Finally, type (1) samples do not show any significant influence of hopping within the measured temperature range. The difference is due to the defect density – higher defect density means higher μ_D and thus a more significant influence of hopping. In the following, only the high-temperature part where the charge carrier concentration is dominated by band transport

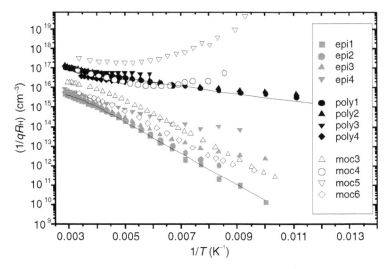

Fig. 7.19. Temperature-dependent charge carrier concentration for various CuGaSe$_2$ layers. Samples labeled *epi...* and *poly...* are epitaxial and polycrystalline samples prepared by PVD; samples labeled *moc...* are epitaxial samples grown by MOVPE [93]

is considered since this transport mechanism allows the determination of defect ionization energies and concentrations. A clear tendency is already seen without fitting the data – with higher free carrier concentration the slope is shallower, i.e., the acceptor activation energy is smaller.

Fits to those parts of the $p(T)$ curves where hopping conduction is not influential are performed according to the solution of the quadratic equation discussed in Sect. 7.2.2:

$$p = -A + \sqrt{A^2 + \frac{1}{g}N_V(N_A - N_D)\ \exp(-E_A/kT)}$$
$$\text{with }\ A = \frac{1}{2}\left(N_D + \frac{1}{g}N_V\ \exp(-E_A kT)\right) \tag{7.19}$$

from which defect ionization energy and defect densities E_A, N_A and N_D are obtained as fit parameters.

Defect ionization energies varying from 10 to 140 meV are obtained, although these samples show the same PL spectra as the samples grown under Cu-excess, depicted in Fig. 7.5. The reason for this large variation is the dependence of the thermal activation energy of the defects on the defect density as shown in Eq. 7.15 and Fig. 7.20. The infinite dilution limit of the defect ionization energy obtained from a fit to Eq. 7.15 is 147 meV. The slope α is obtained as $4.3 \pm 0.3 \times 10^{-8}$ ev cm, in the range obtained for other semiconductors as well [35, 36].

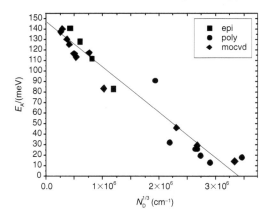

Fig. 7.20. Acceptor activation energy as a function of donor concentration [93]

Thus, the fits to the temperature-dependent charge carrier concentrations result in an acceptor with an ionization energy of $150 \pm 5\,\mathrm{meV}$. This appears reduced due to shielding effects with increasing defect concentration. It should be noted that we have used the donor concentration, i.e., the minority defect concentration for this description, in accordance with the observations of Monecke et al. [37]. Due to the fact that the degree of compensation in these samples is close to 1 (see next paragraph), the plot would yield the same result if the defect activation energy was plotted against the third root of the acceptor concentration.

Self-Compensation

The analysis of the temperature-dependent charge carrier concentration also yields the concentrations of acceptors and donors. Figure 7.21 shows the degree of compensation $K = N_\mathrm{D}/N_\mathrm{A}$ as a function of the acceptor concentration. The degree of compensation is seen to increase with the acceptor concentration [93]. This is a clear signature of self-compensation. Basically, self-compensation describes the effect that when the Fermi level shifts down, i.e., more acceptors are created, the formation enthalpy of the donors decreases. In a simple picture, this can be understood when taking into account the fact that donors are positively charged – if the Fermi-level decreases, i.e., the electron reservoir is reduced, it becomes easier to form positively charged defects. Thus, more donors will form increasing the degree of compensation. Such a behavior can be described by the Fermi-level stabilization model [127–129] and is usually observed as an off-leveling of charge carrier concentration with increasing dopant concentration or by the fact that semiconductors can be doped only p-type or n-type [130, 131]. The effect has been observed till now as the self-compensation of extrinsic dopants by native defects. Here, the

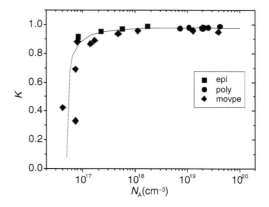

Fig. 7.21. Increasing degree of compensation with increasing acceptor concentration, *the line* is a guide to the eye [93]

effect is for the first time observed as the self-compensation of native defects by native defects.

The effect is observed over the samples at large, independent of their crystallinity or deposition method. The concentration of acceptors depends on the detailed preparation conditions, in particular the Cu-excess. As a tendency, it is observed that the lower defect concentrations occur with higher Cu-excess [132]. Under preparation conditions that favor high acceptor concentrations, the Fermi level would shift towards the valence band. But this decreases the formation energy of the donors and thus leads to increased formation of donors, i.e., increased degree of compensation.

7.5 Defect and Recombination Model for CuGaSe$_2$ and CuInSe$_2$

In this section, a comparison between Hall and PL measurements is made and final conclusions are drawn.

7.5.1 Comparison between PL and Hall Measurements

At a first glance, two discrepancies occur between the PL and Hall measurements discussed here for CuGaSe$_2$: (1) PL shows three acceptors, 60, 100, and 135 meV deep, whereas Hall measurements evidence only one acceptor, 150 meV deep; (2) Hall measurements indicate a high degree of compensation, whereas PL spectra are sharp without the indication of fluctuating potentials that come along with high compensation.

To resolve the second issue, one needs to take into account that fluctuating potentials occur under two conditions – high degree of compensation

and statistical distribution of the defects. It is this statistical distribution
that accounts for the regions with an excess of (positively charged) donors
and the regions with an excess of (negatively charged) acceptors, which cause
the fluctuating electrostatic potentials. Therefore, it must be concluded that
the donors and acceptors that are observed as the DA transitions in the PL
spectra of $CuGaSe_2$, grown under Cu-excess are *not* statistically distributed.
If the donor that is the partner in the DA transition is always created with a
fixed distance in the vicinity of the acceptor, then the different regions with
excess of either donors or acceptors will not occur and thus there will be no
fluctuating potentials. Thus, with *coupled* donor-acceptor pairs, the observed
high degree of compensation can occur without causing fluctuating potentials,
i.e., without distorting the PL spectra.

The deepest acceptor observed in PL has been determined to be $135 \pm$
$10\,meV$, while from Hall measurements an acceptor with an ionization energy
in the infinite dilution limit of $150 \pm 5\,meV$ has been concluded. Within error
margins these two values are equal. Thus, Hall measurements show only the
deepest acceptor. The fact that the acceptors with lower ionization energy, observed in PL are not detected from the Hall measurements is due to the high
compensation. It is usually only the deeper acceptor that can be observed in
Hall measurements in the presence of compensating donors, as one can easily
see for oneself by simulating the solution to Eq. 7.11 for different conditions.
Even if the concentration of the lower-energy acceptors is higher than that of
the deeper acceptor, the $p(T)$ curve will be dominated by the deeper acceptor,
if there are compensating donors of a concentration higher than the concentration of the shallower acceptors. Only at very high temperatures, where the
deeper acceptors are saturated, will the shallower acceptors show up in the
$p(T)$ curve. Thus, it is no contradiction between the two measurements that
three acceptors appear in the PL spectra while only one shows up in the Hall
measurements. The shallower ones are compensated by donors.

7.5.2 Defects in CuGaSe₂ and CuInSe₂

From the discussion of the PL measurements of $CuGaSe_2$ and $CuInSe_2$, it
becomes clear that the shallow defects are the same in both materials – one
donor and three acceptors, with the same characteristic composition dependence. The defect energies in $CuInSe_2$ are somewhat smaller (40 and $60\,meV$
for the dominating acceptors) than in $CuGaSe_2$ (60, 100 and $150\,meV$) as one
would expect from the hydrogen model; taking the effective masses and the
dielectric constants into account, the hydrogen model results in an ionization
energy of $110\,meV$ for acceptors and $16\,meV$ for donors in $CuGaSe_2$, and of
$50\,meV$ for acceptors and $6\,meV$ for donors in $CuInSe_2$. The experimental values differ from those obtained from the hydrogen model, since the dielectric
constant to be taken into account would be a microscopic one, describing the
local environment of the given defect. But the trends are clearly reflected.

Hall measurements in $CuGaSe_2$ show only the deepest acceptor at $150\,meV$ due to compensation of the shallower acceptors.

From the phase diagram, one would expect to obtain stoichiometric material with a very low concentration of native defects when grown under Cu-excess. But the PL and Hall measurements discussed here show that the defect spectrum and the defect concentration change with Cu-excess, even for nominal stoichiometric material. Therefore, it must be concluded that the chalcopyrite grown under Cu-excess is not really stoichiometric, but approaches stoichiometry with increasing Cu-excess. This is supported by the observation that total PL intensities decrease with increasing Cu-excess and that acceptor concentrations and degree of compensation decrease with increasing Cu-excess.

7.6 Appendix: Effective Masses and Dielectric Constants

As indicated in the table below, no measurements of the effective masses of $CuGaSe_2$ exist, only values estimated from other materials. Nevertheless, these values are widely used in the literature as if they were solid. Measurements of the effective masses are urgently needed. They have been hampered so far by the low mobilities of this material.

$CuInSe_2$	$\varepsilon(\omega=0)$	13.6	[133]	capacitance, $300\,K$
	m_e/m_0	0.09	[134]	Faraday effect, $300\,K$
	m_e/m_0	0.08	[135]	cyclotron resonance, $4\,K$
	m_h/m_0	0.71	[136]	optical abs. $300\,K$, heavy hole (light hole 0.09)
	m_h/m_0	0.72	[137]	Burstein–Moss shift
$CuGaSe_2$	$\varepsilon(\omega=0)$	11.0	[138]	IR reflectivity, $300\,K$, $\varepsilon\,(2\varepsilon_\perp + \varepsilon_{||})/3$
	m_e/m_0	0.14	[139]	estimate from other materials
	m_h/m_0	1.2	[89]	estimate from other materials
	m_h/m_0	1.0		used here to fit $p(T)$ data

Acknowledgments

It is a pleasure to express my thanks to everybody in my group for the contribution to this work. Several Ph.D students have done excellent work on this topic: Andreas Bauknecht, Steffen Schuler, and Niklas Rega. I also wish to thank Jürgen Albert and Shiro Nishiwaki for the growth of high-quality epitaxial and polycrystalline films and Yvonne Tomm for her superior single crystals. I owe many colleagues thanks for their support and helpful discussions.

Parts of this work were supported by the European Commission under the Joule Program and the German Ministry of Research (BMBF) in the framework of various projects.

References

1. Mikkelsen, J.C. Jr.: Ternary phase relations of the chalcopyrite compound CuGaSe₂. J. Electron. Mater. 10, 541–558 (1981)
2. Fiechter, S., Tomm, Y., Diesner, K., Weiss, T.: Homogeneity ranges, defect phases and defect formation energies in $A^{I}B^{III}C_2^{VI}$ chalcopyrites (A = Cu; B = Ga, In; C = S, Se). Jpn. J. Appl. Phys. 39 (suppl. 1), 123–126 (1999)
3. Neumann, H.: Vacancy formation enthalpies in $A^{I}B^{III}C_2^{VI}$ chalcopyrite semiconductors. Cryst. Res. Technol. 18, 901–906 (1983)
4. Zunger, A., Zhang, S.B., Wei, S.-H.: Revisiting the defect physics in CuInSe₂ and CuGaSe₂. In: Basore, P. (ed.) 26th IEEE PV Specialist Conference, pp. 313–318. IEEE, New York (2000)
5. Wei, S.-H., Zhang, S.B., Zunger, A.: Effects of Ga addition to CuInSe₂ on its electronic, structural and defect properties. Appl. Phys. Lett. 72, 3199–3201 (1998)
6. Siebentritt, S.: Wide gap chalcopyrites: material properties and solar cells. Thin Solid Films 403–404, 1–8 (2002)
7. Ramanathan, K., Contreras, M.A., Perkins, C.L., Asher, S., Hasoon, F.S., Keane, J., Young, D., Romero, M., Metzger, W., Noufi, R., Ward, J., Duda, A.: Properties of 19.2% efficiency ZnO/CdS/CuInGaSe₂ thin-film solar cells. Prog. Photovolt. Res. Appl. 11, 225–230 (2003)
8. Young, D.L., Keane, J., Duda, A., AbuShama, J.A.M., Perkins, C.L., Romero, M., Noufi, R.: Improved performance in ZnO/CdS/CuGaSe₂ thin-film solar cells. Prog. Photovolt. Res. Appl. 11, 535–541 (2003)
9. Yu, P.Y., Cardona, M.: Fundamentals of Semiconductors. Springer, Berlin Heidelberg New York (1996)
10. Pankove, J.I.: Optical Processes in Semiconductors. Dover Publications, New York (1975)
11. Landsberg, P.T.: Recombination in Semiconductors. Cambridge University Press, Cambridge (1991)
12. Williams, F.: Donor–acceptor pairs in semiconductors. Phys. stat. sol. 25, 493–512 (1968)
13. Bebb, H.B., Williams, E.W.: Photoluminescence: theory. In: Willardson, R.K., Beer, A.C. (eds.) Semiconductors and Semimetals, vol. 8: Transport and Optical Phenomena, pp. 181–320. Academic Press, New York (1972)
14. Dean, P.J.: Inter-impurity recombinations in semiconductors. In: McCaldin, J.O., Somorjai, G. (eds.): Progress in Solid State Chemistry, vol. 8, pp. 1–126. Pergamon Press, Oxford (1973)
15. Schmidt, T., Lischka, K., Zulehner, W.: Excitation-power dependence of the near-band-edge photoluminescence of semiconductors. Phys. Rev. B 45, 8989–8994 (1992)
16. Haynes, J.R.: Experimental proof of the existence of a new complex in silicon. Phys. Rev. Lett. 4, 361–363 (1960)
17. Sharma, R.R., Rodriguez, S.: Theory of excitons bound to ionized impurities in semiconductors. Phys. Rev. 153, 823–827 (1967)
18. Atzmüller, H., Fröschl, F., Schröder, U.: Theory of excitons bound to neutral impurities in polar semiconductors. Phys. Rev. B 19, 3118–3129 (1979)
19. Eagles, D.M.: Optical absorption and recombination radiation in semiconductors due to transitions between hydrogen-like acceptor impurity levels and the conduction band. J. Phys. Chem. Solids 16, 76–83 (1960)

20. Nathan, M.I., Morgan, T.N.: Excitation dependence of photoluminescence in n- and p-type compensated GaAs. In: Kelly, P.L., Lax, B., Tannenwald, P.E. (eds.) Physic of Quantum Electronics, pp. 478–486. McGraw Hill, New York (1966)
21. Maeda, K.: Temperature dependence of pair band luminescence in GaP. J. Phys. Chem. Solids 26, 595–605 (1965)
22. Thomas, D.G., Hopfield, J.J., Augustyniak, W.M.: Kinetics of the radiative recombination of randomly distributed donors and acceptors. Phys. Rev. A 140. 202–220 (1965)
23. Shibata, H.: Negative thermal quenching curves in photoluminescence of solids. Jpn. J. Appl. Phys. 37, 550–553 (1997)
24. Shklovskii, B.I., Efros, A.L.: Electronic Properties of Doped Semiconductors. Springer, Berlin Heidelberg New York (1984)
25. Levanyuk, A.P., Osipov, V.V.: Edge luminescence of direct-gap semiconductors. Sov. Phys. Usp. 24, 187–215 (1981)
26. Gislason, H.P., Yang, B.H., Linnarsson, M.K.: Shifting photoluminescence band in high-resistivity Li-compensated GaAs. Phys. Rev. B 47, 9418–9424 (1993)
27. Bäume, P., Gutowski, J., Kurtz, E., Hommel, D., Landwehr, G.: Intensity-dependent energy and lineshape variation of donor–acceptor-pair bands in highly compensated ZnSe. N. J. Cryst. Growth 159, 252–256
28. Heitz, R., Moll, E., Kutzer, V., Wiesmann, D., Lummer, B., Hoffmann, A., Broser, I., Bäume, P., Taudt, W., Söllner, J., Heuken, M.: Influence of compensation on the luminescence of nitrogen-doped ZnSe epilayers grown by MOCVD. J. Cryst. Growth 159, 307–311 (1996)
29. Seeger, K.: Semiconductor Physics. Springer, Berlin Heidelberg New York (2002)
30. Blakemore, J.S.: Semiconductor Statistics. Pergamon Press, Oxford (1962)
31. Wiley, J.D.: Mobility of holes in III–V compounds. In: Willardson, R.K., Beer, A.C. (eds.) Semiconductors and Semimetals—Transport Phenomena, vol. 10, pp. 91–174. Academic Press, New York (1975)
32. Pearson, G.L., Bardeen, J.: Electrical properties of pure silicon and alloys containing boron and phosphorus. Phys. Rev. 75, 865–883 (1949)
33. Lee, T.F., McGill, T.C.: Variation of impurity-to-band activation energies with impurity density. J. Appl. Phys. 46, 373–380 (1975)
34. Neumark, G.F.: Concentration and temperature dependence of impurity-to-band activation energies. Phys. Rev. B 5, 408–417 (1972)
35. Podör, B.: On the concentration dependence of the thermal ionisation energy of impurities in InP. Semicond. Sci. Technol. 2, 177–178 (1987)
36. Podör, B.: On the concentration dependence of the thermal activation energy of impurities in semiconductors. In: Ferenczi, G., Beleznay, F. (eds.) New Developments in Semiconductor Physics, pp. 55–60. Springer, Berlin Heidelberg New York (1988)
37. Monecke, J., Siegel, W., Ziegler, E., Kühnel, G.: On the concentration dependence of the thermal impurity-to-band activation energies in semiconductors. Phys. stat. sol. (b) 103, 269–279 (1981)
38. Mott, N.F., Davis, E.A.: Electronic Processes in Non-crystalline Materials. Clarendon, Oxford (1971)
39. Holstein, T.: Sign of the Hall coefficient in hopping-type charge-transport. Philosophical Magazine 27 (1), 225–233 (1973)

40. Klinger, M.I.: On the Hall effect in hopping transport. Phys. stat. sol. 31, 545–555 (1969)
41. Movaghar, B., Pohlmann, B., Würtz, D.: The Hall mobility in hopping conduction. J. Phys. C: Solid State Phys. 14, 5127–5137 (1981)
42. Mott, N.F.: Metal–Insulator Transitions. Taylor & Francis, London (1974)
43. Shay, J.L., Tell, B., Kasper, H.M., Schiavone, L.M.: Electronic Structure of AgInSe$_2$ and CuInSe$_2$. Phys. Rev. B 7, 4485–4490 (1973)
44. Shay, J.L., Wagner, S., Kasper, H.M.: Efficient CuInSe$_2$/CdS solar cells. Appl. Phys. Lett. 27, 89–90 (1975)
45. Migliorato, P., Shay, J.L., Kasper, H.M., Wagner, S.: Analysis of the electrical and luminescent properties of CuInSe$_2$. J. Appl. Phys. 46, 1777–1782 (1975)
46. Tell, B., Shay, J.L., Kasper, H.M.: Room-temperature electrical properties of ten I–III–VI$_2$ semiconductors. J. Appl. Phys. 43, 2469 (1972)
47. Parkes, J., Tomlinson, R.D., Hampshire, M.J.: Electrical properties of CuInSe$_2$ single crystals. Solid State Electron. 16, 773–777 (1973)
48. Yu, P.W.: Radiative recombination in melt-grown and Cd-implanted CuInSe$_2$. J. Appl. Phys. 47, 677–684 (1976)
49. Rincón, C., Gonzalez, J.: Influence of impurities on the optical properties of CuInSe$_2$ near the fundamental absorption edge. Phys. stat. sol. (b) 110, K171–K174 (1982)
50. Rincón, C., Gonzalez, J., Sanchez Perez, G.: Luminescence and impurity states in CuInSe$_2$. J. Appl. Phys. 54, 6634–6636 (1983)
51. Massé, G.: Shallow centers in some photovoltaic Cu–III–VI$_2$ compounds. J. Phys. Chem. Solids 45, 1091–1097 (1984)
52. Massé, G., Redjai, E.: Radiative recombination and shallow centers in CuInSe$_2$. J. Appl. Phys. 56, 1154–1159 (1984)
53. Massé, G., Redjai, E.: Lattice defects in I–III–VI$_2$ compounds. J. Phys. Chem. Solids 47, 99–104 (1986)
54. Rincón, C., Bellabarba, C.: Optical properties of copper indium diselenide near the fundamental absorption edge. Phys. Rev. B 33, 7160–7163 (1986)
55. Massé, G.: Concerning lattice defects and defect levels in CuInSe$_2$ and the I–III–VI$_2$ compounds. J. Appl. Phys. 68, 2206–2210 (1990)
56. Dagan, G., Abou-Elfotouh, F.A., Dunlavy, D.J., Matson, J.R., Cahen, D.: Defect level identification in CuInSe$_2$ from photoluminescence studies. Chem. Mater. 2, 286–293 (1990)
57. Massé, G., Djessas, K., Guastavino, F.: Study of CuGa$_x$In$_{1-x}$Se$_2$ and CuGa$_x$In$_{1-x}$Te$_2$ compounds. J. Phys. Chem. Solids 52, 999–1004 (1991)
58. Niki, S., Makita, Y., Yamada, A., Obara, A., Misawa, S., Igarashi, O., Aoki, K., Kutsuwada, N.: Sharp optical emission from CuInSe$_2$ thin films grown by molecular beam epitaxy. Jpn. J. Appl. Phys. 33, L500–L502 (1994)
59. Niki, S., Makita, Y., Yamada, A., Obara, A.: Photoluminescence properties of CuInSe$_2$ grown by molecular beam epitaxy. Sol. Energy Mater. Sol. Cells 35, 141–147 (1994)
60. Niki, S., Shibata, H., Fons, P.J., Yamada, A., Obara, A., Makita, Y., Kurafuji, T., Chichibu, S., Nakanishi, H.: Excitonic emissions from CuInSe$_2$ on GaAs(001) grown by molecular beam epitaxy. Appl. Phys. Lett. 67, 1289–1291 (1995)
61. Zott, S., Leo, K., Ruckh, M., Schock, H.W.: Radiative recombination in CuInSe$_2$ thin films. J. Appl. Phys. 82, 356–367 (1997)

62. Niki, S., Suzuki, R., Ishibashi, S., Ohdaira, T., Fons, P.J., Yamada, A., Oyanagi, H.: Identification and control of intrinsic defects in $CuInSe_2$ films. In: Schmid, J., Ossenbrinck, H.A., Helm, P., Ehmann, H., Dunlop, E.D. (eds.) 2nd World Conference on Photovoltaic Solar Energy Conversion, pp 610 (1998)

63. Wagner, M., Dirnstorfer, I., Hofmann, D.M., Lampert, M.D., Karg, F., Meyer, B.K.: Characterization of $CuIn(Ga)Se_2$ thin films I. Cu-rich layers. Phys. stat. sol. (a) 167, 131 (1998)

64. Dirnstorfer, I., Hofmann, D.M., Meister, D., Meyer, B.M., Riedl, W., Karg, F.: Postgrowth thermal treatment of $CuIn(Ga)Se_2$: characterization of doping levels In-rich thin films. J. Appl. Phys. 85, 1423–1428 (1999)

65. Kazmerski, L.L., Ayyagari, M.S., White, F.R., Sanborn, G.A.: Growth and properties of vacuum deposited $CuInSe_2$ thin films. J. Vac. Sci. Technol. 13, 139–144 (1976)

66. Neumann, H., van Nam, N., Höbler, H.-J., Kühn, G.: Electrical properties of n-type $CuInSe_2$ single crystals. Solid State Commun. 25, 899–902 (1978)

67. Neumann, H., Tomlinson, R.D., Nowak, E., Avgerinos, N.: Electrical properties of p-type $CuInSe_2$ single crystals. Phys. stat. sol. (a) 56, K137–K140 (1979)

68. Neumann, H., Nowak, E., Kühn, G.: Impurity states in $CuInSe_2$. Cryst. Res. Technol. 16, 1369–1376 (1981)

69. Neumann, H., Nowak, E.: Influence of substrate orientation on the electrical properties of $CuInSe_2$ epitaxial layers on GaAs substrates. Cryst. Res. Technol. 18, 779–785 (1983)

70. Neumann, H., Nowak, E., Kühn, G., Heise, B.: The electrical properties of $CuInSe_2$ thin films deposited onto CaF_2 substrates. Thin Solid Films 102, 201–208 (1983)

71. Bardeleben, H.Jv.: Selenium self-diffusion study in the $1-3-6_2$ semiconductor: $CuInSe_2$. J. Appl. Phys. 56, 321–326 (1984)

72. Bardeleben, H.Jv. The chemistry of structural defects in $CuInSe_2$. Solar Cells 16, 381–390 (1986)

73. Wasim, S.M.: Transport properties of $CuInSe_2$. Solar Cells 16, 289–316 (1986)

74. Kühn, G., Neumann, H.: $A^I B^{III} C_2^{VI}$-Halbleiter mit Chalcopyritstruktur. Z. Chem. 27, 197–206 (1987)

75. Rincón, C., Wasim, S.M., Ochoa, J.L.: Shallow donors, metallic conductivity, and metal–insulator transition in n-type $CuInSe_2$. Phys. stat. sol. (a) 148, 251–258 (1995)

76. Chathraphorn, S., Yoodee, K., Songpongs, P., Chityuttakan, C., Sayavong, K., Wongmanerod, S., Holtz, P.O.: Photoluminescence of a high quality $CuInSe_2$ single crystal. Jpn. J. Appl. Phys. 37, L269–L271 (1998)

77. Rega, N., Siebentritt, S., Beckers, I., Beckmann, J., Albert, J., Lux-Steiner, M.C.: Defect spectra in epitaxial $CuInSe_2$ grown by MOVPE. Thin Solid Films 431–432, 186–190 (2003)

78. Zhang, S.B., Wei, S.-H., Zunger, A., Katayama-Yoshida, H.: Defect physics of the $CuInSe_2$ chalcopyrite semiconductor. Phys. Rev. B 57, 9642–9656 (1998)

79. Groenink, J.A., Janse, P.H.: A generalized approach to the defect chemistry of ternary compounds. Z. Physik. Chem. N. F. 110, 17–28 (1978)

80. Mudryi, A.V., Bodnar, I.V., Gremenok, V.F., Victorov, I.A., Patuk, A.I., Shakin, I.A.: Free and bound exciton emission in $CuInSe_2$ and $CuGaSe_2$ single crystals. Sol. Energy Mater. Sol. Cells 53, 247–253 (1998)

81. Paorici, C., Romeo, N., Sberveglieri, G., Tarricone, L.: Electroluminescence in $CuGaSe_2$ single crystals. J. Lumin. 15, 101–103 (1977)

82. Vecchi, M.P., Ramos, J., Giriat, W.: Photoluminescence in CuGaSe$_2$. Solid State Electron. 21, 1609–1612 (1978)

83. Massé, G., Lahlou, N., Yamamoto, N.: Edge emission in CuGaSe$_2$. J. Appl. Phys. 51, 4981–4984 (1980)

84. Massé, G., Djessas, K., Yarzhou, L.: Study of CuGa(Se,Te)$_2$ bulk material and thin films. J. Appl. Phys. 74, 1376–1381 (1993)

85. Yamada, A., Fons, P., Niki, S., Shibata, H., Obara, A., Makita, Y., Oyanagi, H.: A shallow state in molecular beam epitaxial grown CuGaSe$_2$ film detectable by 1.62 eV photoluminescence. J. Appl. Phys. 81, 2794–2798 (1997)

86. Shirakata, S., Chichibu, S., Miyake, H., Suigiyama, K.: Optical properties of CuGaSe$_2$ and CuAlSe$_2$ layers epitaxially grown on Cu(In$_{0.04}$Ga$_{0.96}$)Se$_2$ substrates. J. Appl. Phys. 87, 7294–7302 (2000)

87. Yoshino, K., Sugiyama, M., Maruoka, D., Chichibu, S., Komaki, H., Umeda, K., Ikari, T.: Photoluminescence spectra of CuGaSe$_2$ crystals. Physica B 302–303, 357–363 (2001)

88. Schumann, B., Tempel, A., Kühn, G., Neumann, H., Nam, N.V., Hänsel, T.: Structural and electrical properties of CuGaSe$_2$ thin films on GaAs substrates. Kristall. Technik. 13, 1285–1295 (1978)

89. Mandel, L., Tomlinson, R.D., Hampshire, M.J., Neumann, H.: Electrical properties of CuGaSe$_2$ single crystals. Solid State Commun. 32, 201–204 (1979)

90. Stankiewicz, J., Giriat, W., Ramos, J., Vecchi, M.P.: Electrical and optical properties of CuGaSe$_2$. Sol. Energy Mater. 1, 369–377 (1979)

91. Schön, J.H., Baumgartner, F.P., Arushanov, E., Riazi-Nejad, H., Kloc, C., Bucher, E.: Potoluminescence and electrical properties of Sn-doped CuGaSe$_2$ single crystals. J. Appl. Phys. 79, 6961–6965 (1996)

92. Bauknecht, A., Siebentritt, S., Albert, J., Lux-Steiner, M.C.: Radiative recombination via intrinsic defects in CuGaSe$_2$. J. Appl. Phys. 89, 4391–4400 (2001)

93. Schuler, S., Siebentritt, S., Nishiwaki, S., Rega, N., Beckmann, J., Brehme, S., Lux-Steiner, M.C.: Self-compensation of intrinsic defects in the ternary semiconductor CuGaSe$_2$. Phys. Rev. B 69, 45210 (2004)

94. Siebentritt, S., Schuler, S.: Defects and transport in the wide gap chalcopyrite CuGaSe$_2$. J. Phys. Chem. Solids 64, 1621–1626 (2003)

95. Siebentritt, S., Beckers, I., Riemann, T., Christen, J., Hoffmann, A., Dworzak, M.: Reconciliation of luminescence and Hall measurements on the ternary semiconductor CuGaSe$_2$. Appl. Phys. Lett. 86 (2005) 091909.

96. Tanaka, S., Kawami, S., Kobayashi, H., Sasakura, H.: Luminescence in CuGaS$_{2-2x}$Se$_{2x}$ mixed crystals grown by chemical vapor transport. J. Phys. Chem. Solids 38, 680–681 (1977)

97. Shirakata, S., Chichibu, S., Sudo, R., Ogawa, A., Matsumoto, S., Isomura, S.: Photoreflectance and photoluminescence studies of CuAl$_x$Ga$_{1-x}$Se$_2$ alloys. Jpn. J. Appl. Phys. 32, L1304–L1307 (1993)

98. Chichibu, S., Shirakata, S., Isomura, S., Nakanishi, H.: Visible and ultraviolet photoluminescence from Cu–III–VI$_2$ chalcopyrite semiconductors grown by metalorganic vapor phase epitaxy. Jpn. J. Appl. Phys. 36, 1703–1714 (1997)

99. Schön, J.H., Bucher, E.: Comparison of point defects in CuInSe$_2$ and CuGaSe$_2$ single crystals. Sol. Energy Mater. Sol. Cells 57, 229–237 (1999)

100. Susaki, M., Miyauchi, T., Horinaka, H., Yamamoto, N.: Photoluminescence properties of CuGaSe$_2$ grown by iodine vapour transport. Jpn. J. Appl. Phys. 17, 1555–1559 (1978)

101. Arndt, W., Dittrich, H., Schock, H.W.: CuGaSe$_2$ thin films for photovoltaic applications. Thin Solid Films 130, 209–216 (1985)
102. Sugiyama, K., Iwasaki, S., Endo, T., Miyake, H.: Photoluminescence of CuAl$_x$Ga$_{1-x}$Se$_2$ crystals grown by chemical vapor transport. J. Appl. Phys. 65, 5212–5215
103. Shirakata, S., Tamura, K., Isomura, S.: Molecular beam epitaxy of CuGaSe$_2$ on GaAs Substrate using metalorganic copper and gallium precursors. Jpn. J. Appl. Phys. 35, L531–L534 (1996)
104. Klenk, M., Schenker, O., Bucher, E.: Photoluminescence and X-ray fluorescence measurements of successively thinned CuGaSe$_2$-films. Thin Solid Films 361–362, 229–233 (1999)
105. Rusu, M., Gashin, P., Simashkevich, A.: Electrical and luminescent properties of CuGaSe$_2$ crystals and thin films. Sol. Energy Mater. Sol. Cells 70, 175–186 (2001)
106. Meeder, A., Fuertes-Marrón, D., Tezlevan, V., Arushanov, E., Rumberg, A., Schedel-Niedrig, T., Lux-Steiner, M.C.: Radiative recombination in CVT-grown CuGaSe$_2$ single crystals and thin films. Thin Solid Films 431–432, 214–218 (2003)
107. Siebentritt, S., Rega, N., Zajogin, A., Lux-Steiner, M.C.: Do we really need another photoluminescence study on CuInSe$_2$? Phys. stat. sol. (c) 1, 2304–2310 (2004)
108. Rincón, C., Wasim, S.M.: Defect chemistry of A^{I}B^{III}C$_2^{VI}$ chalcopyrite semiconducting compounds. In: 7th International Conference on Ternary and Multinary Compounds, pp. 443–451. MRS, Pittsburgh (1987)
109. Wagner, M., Hofmann, D.M., Dirnstorfer, I., Lampert, M.D., Karg, F., Meyer, B.K.: Characterization of CuIn(Ga)Se$_2$ thin films II. Magneto-optical properties of Cu- and In-rich layers. Phys. stat. sol. (a) 168, 153 (1998)
110. Dirnstorfer, I., Wagner, M., Hofmann, D.M., Lampert, M.D., Karg, F., Meyer, B.K.: Characterization of CuIn(Ga)Se$_2$ thin films III. In-rich layers. Phys. stat. sol. (a) 168, 163 (1998)
111. Dirnstorfer, I., Burkhardt, W., Kriegseis, W., Österreicher, I., Alves, H., Hofmann, D.M., Ka, O., Polity, A., Meyer, B.K., Braunger, D.: Annealing studies on CuIn(Ga)Se$_2$: the influence of gallium. Thin Solid Films 361/362, 400–405 (2000)
112. Krustok, J., Collan, H., Yakushev, M., Hjelt, K.: The role of spatial potential fluctuations in the shape of the PL bands of multinary semiconductor compounds. Phys. Scripta T 79, 179–182 (1999)
113. Mudryi, A.V., Bodnar, I.V., Victorov, I.A., Gremenok, V.F., Patuk, A.I., Shakin, I.A.: Characterization by photoluminescence of CuGa$_x$In$_{1-x}$Se$_2$ thin films prepared by laser deposition. Inst. Phys. Conf. Ser. 152, 465–468 (1997)
114. Yoshino, K., Yokoyama, H., Maeda, K., Ikari, T.: Crystal growth and photoluminescence of CuIn$_x$Ga$_{1-x}$Se$_2$ alloys. J. Cryst. Growth 211, 476–479 (2000)
115. Keyes, B.M., Dippo, P., Metzger, W., Abushama, J., Noufi, R.: Cu(In,Ga)Se$_2$ thin film evolution during growth—a photoluminescence study. In: IEEE Photovoltaic Specialist Conference, pp. 511–514. IEEE, New York (2002)
116. Schroeder, D.J., Hernandez, J.L., Berry, G.D., Rockett, A.A.: Hole transport and doping states in epitaxial CuIn$_{1-x}$Ga$_x$Se$_2$. J. Appl. Phys. 83, 1519–1526 (1998)

117. Chichibu, S., Harada, Y., Uchida, M., Wakiyama, T., Matsumoto, S., Shirakata, S., Isomura, S., Higuchi, H.: Heteroepitaxy and characterization of CuGaSe$_2$ layers grown by low-pressure metalorganic chemical-vapor depostion. J. Appl. Phys. 76, 3009–3015 (1994)
118. Artaud-Gillet, M.C., Duchemin, S., Odedra, R., Orsal, G., Rega, N., Rushworth, S., Siebentritt, S.: Evaluation of copper organometallic sources for CuGaSe$_2$ photovoltaic applications. J. Cryst. Growth 248, 163–168 (2003)
119. Rega, N., Siebentritt, S., Beckers, I., Beckmann, J., Albert, J., Lux-Steiner, M.C.: MOVPE of epitaxial CuInSe$_2$ on GaAs. J. Cryst. Growth 248, 169–174 (2003)
120. Boehnke, U.-C., Kühn, G.: Phase relations in the ternary system Cu–In–Se. J. Mater. Sci. 22, 1635–1641 (1987)
121. Mickelsen, R.A., Chen, W.S.: High photocurrent polycrystalline thin-film CdS/CuInSe$_2$ solar cell. Appl. Phys. Lett. 36, 371–373 (1980)
122. Nishiwaki, S., Ennaoui, A., Schuler, S., Siebentritt, S., Lux-Steiner, M.C.: Surface treatments and properties of CuGaSe$_2$ thin films for solar cell application. Thin Solid Films 431–432, 296–300 (2003)
123. Bauknecht, A., Siebentritt, S., Albert, J., Tomm, Y., Lux-Steiner, M.-C.: Excitonic photoluminescence from CuGaSe$_2$ single crystals and epitaxial layers: temperature dependence of the band gap energy. Jpn. J. Appl. Phys. 39 (Suppl. 1), 322–325 (2000)
124. Ka, O., Yamada, A.: Excited-state spectroscopy of an effective-mass-like level in Cu-rich CuGaSe$_2$. Thin Solid Films 361–362, 509–513 (2000)
125. Fearheiley, M.: The phase relations in the Cu, In, Se system and the growth of CuInSe$_2$ single crystals. Solar Cells 16, 91–100 (1986)
126. Tanino, H., Maeda, T., Fujikake, H., Nakanishi, H.Y., Endo, S., Irie, T.: Raman spectra of CuInSe$_2$. Phys. Rev. B 45, 13323–13330 (1992)
127. Walukiewicz, W.: Fermi level dependent native defect formation: consequences for metal–semiconductor and semiconductor–semiconductor interfaces. J. Vac. Sci. Technol. B 6, 1257–1262 (1988)
128. Walukiewicz, W.: Mechanism of Fermi-level stabilization in semiconductors. Phys. Rev. B 37, 4760–4763 (1988)
129. Walukiewicz, W.: Defects and self-compensation in semiconductors, In Siebentritt, S., Rau, U. (Eds.) Wide-gap chalcopyrites, pp. 33–50, Springer, Berlin (2005)
130. Agrinskaya, N.V., Mashovets, T.V.: Self-compensation in semiconductors: a review dedicated to the hundredth anniversary of the birthday of Yakov Il'ich Frenkel. Semiconductors 28, 843–857 (1994)
131. Desnica, U.V.: Doping limits in II–VI compounds—challenges, problems and solutions. Prog. Cryst. Growth Charact. 36, 291–357 (1998)
132. Siebentritt, S., Gerhard, A., Brehme, S., Lux-Steiner, M.: Composition dependent doping and transport properties of CuGaSe$_2$. Mater. Res. Soc. Symp. Proc. 668, H.4.4.1–H.4.4.6 (2001)
133. Li, P.W., Anderson, R.A., Plovnick, R.H.: Dielectric constant of CuInSe$_2$ by capacitance measurements. J. Phys. Chem. Solids 40, 333–334 (1979)
134. Weinert, H., Neumann, H., Höbler, H.-J., Kühn, G., van Nam, N.: Infrared Faraday effect in n-type CuInSe$_2$. Phys. stat. sol. (b) 81, K59–K61 (1977)
135. Arushanov, E., Essaleh, L., Galibert, J., Leotin, J., Askenazy, S.: Shubnikov–De Haas oscillations in CuInSe$_2$. Physica B 184, 229–231 (1993)

136. Neumann, H., Sobotta, H., Kissinger, W., Riede, V., Kühn, G.: Hole effective masses in CuInSe$_2$. Phys. stat. sol. (b) 108, 483–487 (1981)
137. Chattopadhyay, K.K., Sanyal, I., Chaudhuri, S., Pal, A.K.: Burstein–Moss shift in CuInSe$_2$ films. Vacuum 42, 915 (1991)
138. Syrbu, N.N., Bogdanash, M., Tezlevan, V.E., Mushcutariu, I.: Lattice vibrations in CuIn$_{1-x}$Ga$_x$Se$_2$ crystals. Physica B 229, 199–212 (1997)
139. Quintero, M., Rincón, C., Grima, P.: Temperature variation of energy gaps and deformation potentials in CuGa(S$_z$Se$_{1-z}$)$_2$ semiconductor alloys. J. Appl. Phys. 65, 2739–2743 (1989)

8

Spatial Inhomogeneities of Cu(InGa)Se$_2$ in the Mesoscopic Scale

G.H. Bauer

8.1 Introduction

Among thin-film semiconductors presently considered most promising for photovoltaic applications, Cu(InGa)Se$_2$ (CIGS) in terms of efficiency of laboratory cells and of modules from a pilot line is the favorite candidate [1–3]. Despite the granularity of CIGS with grain sizes in the μm or sub-μm range, the collection of photoexcited charges (short-circuit current) seems not to be substantially affected by grain boundaries. The open-circuit voltage V_{oc} shows the deficit to the theoretical upper limit [4] common for all thin film solar cells in which about 60% of the theoretical maximum V_{oc} is achieved.

The presence of Na during film formation is known to have an extremely beneficial influence on crystallographic properties and on grain boundary properties of polycrystalline CIGS as well [5], and the efficiencies of polycrystalline solar cells exceed those with epitaxially grown single-crystalline absorbers.

From structural analyses, such as various types of X-ray scattering, atomic force microscopy (AFM) scans, average grain sizes and their distribution, the orientation of grains, structural defects and voids have been detected. However, only a few analyses of optoelectronic properties, in particular those that significantly govern the photovoltaic function with a spatial resolution in the lengths scale of the grain sizes or below are available [6–8]; consequently no consistent distinction of the influence of spatial inhomogeneities, and corresponding entropic terms responsible for a further drop of solar cell efficiencies have been derived.

Published data extracted from traditional as well as sophisticated experiments, e.g., deep-level transient spectroscopy (DLTS), admittance and capacitance spectroscopy under different external conditions, optical transmission/reflection/absorption, quantum yield and current–voltage analyses, of course, have not been performed with a lateral resolution of a micrometer or below.

8.2 Micron and Sub-Micron Resolution Methods

For structural information, particularly for the surface topology function $z = z(x, y)$, several AFM scans with a cantilever apparatus with a lateral resolution of $\Delta x = \Delta y \leq 100$ nm (depth resolution $\Delta z \leq 5$ nm) of different scan regimes $[(2\,\mu m)^2$ to $(200\,\mu m)^2]$ and at different positions of the samples have been performed.

In a confocal microscope setup, outlined in Fig. 8.1, we have simultaneously (at the very same position) monitored lateral photoluminescence (PL) and optical reflection from the respective absorbers with excitation at $\lambda = 632.8$ nm in order to compare topological features such as grains and their boundaries with the lateral luminescence that yields the quality of the photoexcitation state. The lateral resolution of both signals is close to the diffraction limit and amounts to about $\Delta x = \Delta y \approx 0.85\,\mu m$ for the reflection ($\lambda = 633$ nm) and $\Delta x = \Delta y \approx 0.9\,\mu m$ for the PL yield at $\lambda \approx 1.0\,\mu m$ [6].

After a translation of the spatially recorded topology $z = z(x\,y)$, the optical reflection and the luminescence via either 2-D Fourier transforms or Minkowski-erosionn–dilatation operations, we have checked for the necessary minimum size of scan regimes to be representative of the respective data sets.

8.3 Samples

The set of samples consists of $Cu(In_{1-x}Ga_x)Se_2$ absorbers prepared by Z.S.W. Stuttgart under conditions very similar to that of a pilot line for CIGS

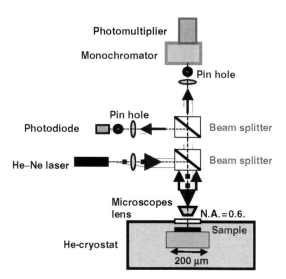

Fig. 8.1. Confocal microscope setup for simultaneous detection of luminescence and optical reflection with sub-micron lateral resolution

modules fabrication [9–11] of nominally 15% efficiency for $x = 0.3$. Here, absorbers with varying Ga concentration [12] in the range of $0.1 \leq x \leq 1.0$ have been studied with thicknesses of about 2–2.5 μm, deposited on glass substrates with and without Mo rear contacts and commonly coated with a 50 nm thick CdS layer from chemical bath deposition [9], forming the heterojunction and simultaneously serving as a surface passivation layer that prevents oxidation and penetration of water into the CIGS and avoids corresponding irreversible changes of optoelectronic properties.

8.4 Analyses

Due to the variations of AFM signals in lateral (x, y) and z-directions on length scales of at least several tens of nm, we may regard the function z_{AFM} as topological surface contour functions $z = z(x, y)$ almost exclusively induced by the polycrystallinity of the CIGS layer.

The response of the optical reflection that may be interpreted as some qualitative picture of the surface topology has been simultaneously recorded with the luminescence. It consists of that part of photons of the incident light spot with approximately radial Gaussian shape ($\Gamma_0 = \Gamma_{00} \exp[-a^2(x^2+y^2)]$) that is reflected at the surface with the contour function $z = z(x, y)$ and according to the contrast in refractive indices air/CIGS, which is subsequently accepted by the aperture of the confocal microscope and transmitted to the detector diode (see Fig. 8.1). The acceptance angle α_c of the microscope for reflected photons amounts in air to $\alpha_c = \arctan(\Phi_m/2\xi_0)$, which translates into an acceptance angle for photons traveling in CIGS of $\alpha_c^* = (n_{\mathrm{air}}/n_{\mathrm{CIGS}}) \arctan(\Phi_m/2\xi_0)$ (Fig. 8.2). The step-like acceptance function is $S(\alpha, \alpha_c, r(\omega)) = 0$ for $\alpha > \alpha_c$ and $S(\alpha, \alpha_c, r(\omega)) = r(\omega)$ for $\alpha \leq \alpha_c$.

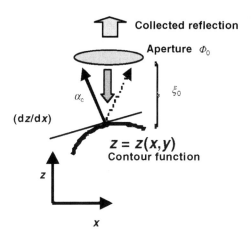

Fig. 8.2. Schematic collection/acceptance by aperture of photons reflected from the surface with respective contour function $z = z(x, y)$

The signal recorded from the optical reflection Γ_r reads analytically [7]:

$$\Gamma_\gamma = \int\limits_{-x_0}^{x_0} \int\limits_{-y_0}^{y_0} \Gamma_{00} \exp\left[-a^2(x^2+y^2)\right] S\left(\alpha, \alpha_c, r(\omega)\right) \, dx \, dy. \qquad (8.1)$$

As a consequence of the detection of intensity signals (squares of wave amplitudes) the reflection signal cannot be unambiguously attributed to either convex or concave contour functions and thus by high reflection response from the plateaus of the tops and the bottom of topological ripples, one spatial period of the contour may create two spatial periods in the reflection signal, i.e., the spatial frequency of those surface structures in the reflection response are doubled (Fig. 8.3).

Moreover, for sufficiently steep contours, e.g., inverted U-type cross-sections, photons that are reflected particularly from the neighborhood of the boundaries of small grains do not reach the aperture and thus the optical reflection will additionally project smaller features and accordingly higher spatial frequencies than the AFM topology signals (see Fig. 8.2.).

The luminescence emitted by matter in terms of the spectral photon flux density $\Gamma(\omega)$ may be formulated by Planck's generalized law [13–15]

$$\Gamma_{\mathrm{Pl}}(\omega) = E(\omega)\omega^2 \left[\exp\left(\frac{\hbar\omega - \mu_{\mathrm{phot}}}{kT}\right) - 1\right]^{-1}, \qquad (8.2)$$

provided one-electron description and quasi-Fermi approach are valid. In Eq. 8.2, the quantities $E(\omega)$, $\hbar\omega$, kT, and μ_{phot} denote spectral emissivity, photon energy, thermal energy and chemical potential of the photon field.

For exclusively radiative transitions of absorption ($r_{\mathrm{abs}} = r_{12}$), spontaneous ($r_{\mathrm{em,sp\,sp}} = r_{21,\mathrm{sp}}$) and stimulated emission ($r_{\mathrm{em,st}} = r_{21,\mathrm{st}}$) in an

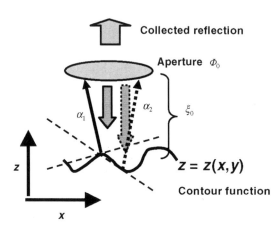

Fig. 8.3. Schematic representation of ambiguity of reflection response of a surface with wavy structure

electronic two-band system with energy levels/energy bands $\varepsilon_2 > \varepsilon_1$, and densities of states $D_1 = D(\varepsilon_1)$, $D_2 = D(\varepsilon_2)$, we introduce the thermal non-equilibrium occupation of electrons in states at ε_2, and the respective hole occupation at ε_1 – the electron densities at ε_1 and the hole concentration at ε_2 are determined by charge/spin conservation – and write the rate equations in steady-state by a rate

$$r_{\text{abs}} = \Gamma n\left(\varepsilon_1\right) p\left(\varepsilon_2\right) B_{12} = r_{\text{em,sp}} + r_{\text{em,st}} = A_{21} n\left(\varepsilon_2\right) p\left(\varepsilon_1\right) + \Gamma B_{12} n\left(\varepsilon_2\right) p\left(\varepsilon_1\right)$$

(8.3)

with $n\left(\varepsilon_2\right) = D(\varepsilon_2) f_{\text{F,n}}$, $p\left(\varepsilon_1\right) = D(\varepsilon_1) f_{\text{F,p}}$, and the quasi-Fermi distribution functions

$$f_{\text{F,n}}(\varepsilon_2) = \left[\exp\left(\frac{\varepsilon_2 - \varepsilon_{\text{F,n}}}{kT}\right) + 1\right]^{-1}$$

(8.4a)

for electrons and

$$f_{\text{F,p}}(\varepsilon_1) = \left[\exp\left(\frac{\varepsilon_{\text{F,p}} - \varepsilon_1}{kT}\right) + 1\right]^{-1}$$

(8.4b)

for holes. Accordingly, it holds that $n(\varepsilon_1) = D(\varepsilon_1)[1 - f_{\text{F,p}}(\varepsilon_1)]$ and $p(\varepsilon_2) = D(\varepsilon_2)[1 - f_{\text{F,n}}(\varepsilon_2)]$.

From the above balance of the rates, we determine the resulting steady-state (PL) photon flux:

$$\Gamma_{\text{pl}} \propto \left[\exp\left(\frac{\hbar\omega - (\varepsilon_{\text{F,n}} - \varepsilon_{\text{F,p}})}{kT}\right) - 1\right]^{-1},$$

(8.5)

where the difference $(\varepsilon_{\text{F,n}} - \varepsilon_{\text{F,p}})$ between the quasi-Fermi levels $\varepsilon_{\text{F,n}}$ and $\varepsilon_{\text{F,p}}$ for electrons and holes equals the chemical potential $\mu_{\text{n,p}}$ of the electron–hole ensemble, that in turn, equals the chemical potential μ_{phot} of the photon field. Thus, we have finally

$$\varepsilon_{\text{F,n}} - \varepsilon_{\text{F,p}} = \mu_{\text{n,p}} = \mu_{\text{phot}}.$$

(8.6)

By quantitative analysis of the luminescence yield and with known spectral emissivity $E(\omega)$, which commonly is derived through spectral absorption measurements considering the energy dependent absorption coefficient and thickness, we get experimental access to the splitting of the quasi-Fermi levels in the respective photoexcited matter. Since in a final device, e.g., a solar cell under illumination, the open-circuit voltage V_{oc} cannot exceed the splitting of the quasi-Fermi levels, we directly see the upper limit of V_{oc} by

$$eV_{\text{oc}} \leq \varepsilon_{\text{F,n}} - \varepsilon_{\text{F,p}} = \mu_{\text{n,p}} = \mu_{\text{phot}}$$

(8.7)

and we are even able to predict it.

8.5 Experimental Results with Lateral Sub-Micro Resolution

Figure 8.4 provides an overview of the topological properties of CIGS absorbers by side views on AFM contours, and the respective roughness RMS data are shown in Fig. 8.5. The overcoating 50 nm-thick CdS window layer is assumed to modify the surface topology only negligibly. The pronounced minimum of the roughness at about $x = 0.3$ is not necessarily an intrinsic structural property of CIGS, but may result from the fact that CIGS of this composition is by far the best known and most optimized quaternary absorber.

In optical reflection we first recognize that about 3–7% of the incident photons are reflected and accepted by the confocal aperture. The linear extrapolation from the solid angle of the aperture (0.89) to the maximum angle of 2π would cause reflection factors of less than 15–40%, indicating that greater than 85–60% of the photons for the excitation are coupled in the absorber. The optical reflection reflects only qualitative features of the surface

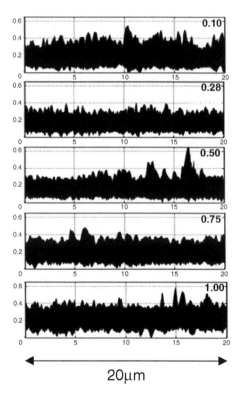

20μm

Fig. 8.4. Side view of AFM topologies of $Cu(In_{1-x}Ga_x)Se_2$ absorbers for different Ga concentrations x. Peaks at the *top* represent the interface to air, peaks at the *bottom* represent "end" of voids and the transition to quasi-compact bulk material

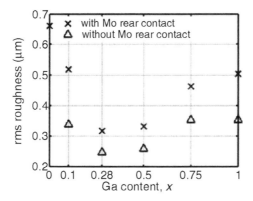

Fig. 8.5. Average roughness data (RMS) from 2-dimensional AFM–topology scans of Cu(In$_{1-x}$Ga$_x$)Se$_2$ absorbers vs Ga concentration

topology and thus does not provide for an estimation of the spatial frequency distribution after Fourier transforms; nevertheless, since it has been recorded simultaneously with the luminescence, we may correlate spatial features of both magnitudes, topology, obviously governed by grains and luminescence, strongly dependent on the local splitting of quasi-Fermi levels (see Fig. 8.6). Surprisingly, the lateral quality of the photoexcited state ($\varepsilon_{F,n} - \varepsilon_{F,p}$) does not show the features of the grains but much larger spatial structures, even when considering the factor of $2\sqrt{2}$ that is introduced by the ambiguity of the reflection response with respect to topological wavy structures.

For the explanation of the variation of PL yields by factors of up to more than 10 [6,16,17], we may certainly exclude the effect of the lateral variation of the amount of laser excitation photons coupled into the absorber, which is in the range of 80–90%; the lion's share of the comparatively large PL

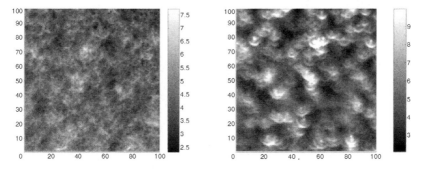

Fig. 8.6. Lateral scans $(100\,\mu\text{m})^2$ of optical reflection R^* (*left*, reflection factors accepted by the aperture range from 2.5% to 7.5%) and of PL yield Y_{pl} (*right*, arbitrary units, note the local variation of about a factor of 5) of a Cu(In$_{0.72}$Ga$_{0.28}$)Se$_2$ absorber [glass/CIGS ($2\,\mu$m)/CdS ($50\,$nm)] [6]

fluctuation undoubtedly results from local variations in excess carrier concentration, reflected by splitting of quasi-Fermi levels($\varepsilon_{F,n} - \varepsilon_{F,p}$).

The local regimes with different PL yields, in addition, exhibit particular spectral properties based on the respective local emissivity and consequently attributed to specific combined density of states of the initial and final states of the radiative transitions of which the origin points towards different metallurgical phases of grains, and grain surfaces (Fig. 8.7).

Even in a complete solar cell with Mo rear and ZnO front contact, say with top and bottom equipotential surfaces, lateral variations of PL yields corresponding to at least 25–30 meV lateral fluctuation in ($\varepsilon_{F,n} - \varepsilon_{F,p}$) at 300 K and AM1.5 fluxes have been observed (Fig. 8.8).

Due to the limited sensitivity of confocal recording of luminescence photons from a spot size less than $1\,\mu m^2$ the PL experiments have been performed at lower than room temperature ($T = 80\,K$) and at excitation photon fluxes much higher than the AM1.5 equivalent ($\Phi = 5 \times 10^4 \Phi_{nAM1.5}$).

In steady-state, in our electronic system – consisting of at least three levels/bands – the generation rate g equals the total recombination rate, which

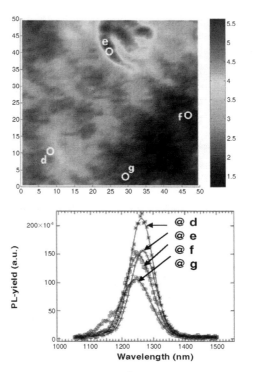

Fig. 8.7. Fig. 8.7. Lateral scan $(50\,\mu m)^2$ of PL yield of a $Cu(In_{0.72}Ga_{0.28})Se_2$ absorber (glass/CIGS $(2\,\mu m)$/CdS $(50\,nm)$) (*upper part*) and spectral shapes of luminescence at different local positions (*lower part*), which indicate a spatial variation in composition (variation in metallurgical phases)

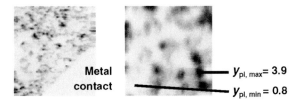

Scan (200 μm)² Scan (40 μm)²

Fig. 8.8. Luminescence yields from a completed CIGS heterodiode (Mo/CIGS/CdS/ZnO). Note that even with boundary conditions of highly conductive front and rear contact surfaces considerable fluctuation of $(\varepsilon_{F,n} - \varepsilon_{F,p})$ occurs

comprises of the radiative rate r_{rad} and each of the i nonradiative transitions according to

$$g = r_{\mathrm{rad}} + \sum_{j=1}^{i} r_{\mathrm{nonrad},i}. \tag{8.8}$$

For the high excitation fluxes used here, we find an almost linear dependence of $r_{\mathrm{rad}} \propto g$, which indicates [extension of Shockley–Reed–Hall (SRH) recombination to extremely high excitation] the paths of nonradiative transitions to be fairly saturated and for further excitation the radiative recombination balances for the increase in generation (see e.g., Fig. 8.9) and we have

$$\sum_{j=1}^{i} r_{\mathrm{nonrad},i} = \mathrm{constant} \neq f(\varPhi) \tag{8.9}$$

$$\text{and } r_{\mathrm{rad}} \gg \sum_{j=1}^{i} r_{\mathrm{nonrad},i} \tag{8.10}$$

The splitting of quasi-Fermi levels $(\varepsilon_{F,n} - \varepsilon_{F,p})$ may, of course, be formulated via the excess carrier concentrations n, p with respect to the appropriate thermal equilibrium densities n_0, p_0 of the initial and final states of the radiative transitions. With Boltzmann approximation we write

$$\varepsilon_{F,n} - \varepsilon_{F,p} = kT \ln\left(\frac{np}{n_0 p_0}\right). \tag{8.11}$$

For different luminescence yields from particular positions being governed by respective differences in $(\varepsilon_{F,n} - \varepsilon_{F,p})$ we get

$$(\varepsilon_{F,n} - \varepsilon_{F,p})_k - (\varepsilon_{F,n} - \varepsilon_{F,p})_l = kT \left[\ln\left(\frac{np}{n_0 p_0}\right)_k - \ln\left(\frac{np}{n_0 p_0}\right)_l\right] \tag{8.12}$$

and moreover,

166   G.H. Bauer

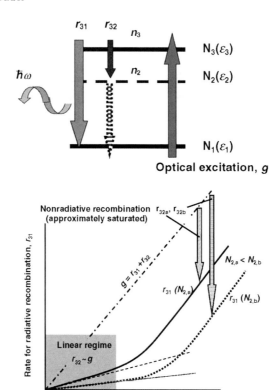

Fig. 8.9. Electronic three-level system with densities of state N_1 (ε_1), N_2 (ε_2), N_3 (ε_3) and the corresponding transition rates r_{13} (optical generation), r_{31} (radiative), and r_{32} (nonradiative) transition (*upper part*). Schematic plot of rates for radiative transitions r_{31} vs generation rate g of an electronic three-level system (extended SRH kinetics) for two defect densities N_{2a}, N_{2b} ($N_{2a} < N_{2b}$) (*lower part*). Nonradiative rates r_{32} (*arrows*) saturate for high generation rates

$$\Delta(\varepsilon_{F,n} - \varepsilon_{F,p})_{kl} = kT \ln\left(\frac{n_k p_k}{n_l p_l}\right) = kT \ln\left(\frac{Y_{pl,k}}{Y_{pl,l}}\right), \qquad (8.13)$$

where $Y_{pl,k}$, $Y_{pl,l}$ designate the luminescence yields at these different local regimes. Here, we see the individual saturated local nonradiative recombination rates

$$r_{nonrad,tot} = \sum_{j=1}^{i} r_{nonrad,i} = g - r_{rad} \qquad (8.14)$$

that are linearly related to the local concentration of defects and their capture cross-section

$$r_{nonrad,k} = N_{defect,k}\sigma_{rec}. \qquad (8.15)$$

For comparably low excitation in the neighborhood of AM1.5 equivalent fluxes, we find from nonlaterally resolved PL studies also a linear behavior of Y_{pl} vs excitation flux Φ that in SRH kinetics can be understood by the excess concentration in the initial state for Y_{pl} to be governed by nonradiative transitions and the corresponding defect density, and the final state concentration of Y_{pl} to be constant (majority carrier related). Accordingly, the radiative rate is proportional to the inverse defect density providing for nonradiative processes:

$$r_{\mathrm{rad}} \propto \frac{1}{N_{\mathrm{defect}}}. \tag{8.16}$$

With these relations we are able to translate the local quasi-Fermi-level splitting created under high excitation into the splitting at low excitation, e.g., at AM1.5 conditions by

$$\Delta(\varepsilon_{\mathrm{F,n}} - \varepsilon_{\mathrm{F,p}})_{\mathrm{kl}}\big|_{\mathrm{lowexc}} = kT \left[\ln\left(\frac{Y_{\mathrm{pl,k}}}{Y_{\mathrm{pl,0}}}\right) - \ln\left(\frac{Y_{\mathrm{pl,l}}}{Y_{\mathrm{pl,0}}}\right) \right] \tag{8.17}$$

and with

$$Y_{\mathrm{pl}}\big|_{\mathrm{lowexc}} \propto r_{\mathrm{rad}}\big|_{\mathrm{lowexc}} = \frac{1}{N_{\mathrm{defect}}} \propto \frac{1}{g - r_{\mathrm{rad}}}\bigg|_{\mathrm{highexc}} \tag{8.18}$$

we finally get

$$\Delta(\varepsilon_{\mathrm{F,n}} - \varepsilon_{\mathrm{F,p}})_{\mathrm{kl}}\big|_{\mathrm{lowexc}} = kT \ln\left(\frac{g - r_{\mathrm{rad,l}}}{g - r_{\mathrm{rad,k}}}\right) = kT \ln\left(1 + \frac{r_{\mathrm{rad,k}} - r_{\mathrm{rad,l}}}{g - r_{\mathrm{rad,k}}}\right). \tag{8.19}$$

Provided the nonradiative rates decrease at least similarly or more rapidly with rising temperature compared to the radiative rates the separation of the quasi-Fermi levels and accordingly the maximum achievable open-circuit voltage can be written as

$$\Delta(eV_{\mathrm{oc}})\big|_{\mathrm{lowexc,300K}} = \Delta(\varepsilon_{\mathrm{F,n}} - \varepsilon_{\mathrm{F,p}})_{\mathrm{kl}}\big|_{\mathrm{lowexc,300K}}$$

$$\geq kT \ln\left(1 + \frac{r_{\mathrm{rad,k}} - r_{\mathrm{rad,l}}}{g - r_{\mathrm{rad,k}}}\right)\bigg|_{\mathrm{highexc,lowTemp.}} \tag{8.20}$$

With the assumption that for an increase in temperature the decay of the nonradiative rates is at least as strong as that of the radiative rates – usually nonradiative transitions are thermally activated and their decay is much stronger – from the above relation for the results in Fig. 8.6 we can estimate

$$\ln\left(1 + \frac{r_{\mathrm{rad,k}} - r_{\mathrm{rad,l}}}{g - r_{\mathrm{rad,k}}}\right)\bigg|_{\mathrm{highexc,lowTemp}} \approx \ln\left[1 + (2\ldots5)\right] \approx (1\ldots1.8) \tag{8.21}$$

the lateral variation of V_{oc} at room temperature and AM1.5 fluxes amount at least to about 26–47 mV.

A further estimation of the variation of the quasi-Fermi levels in CIGS at 1 sun photon fluxes and 300 K has been deduced from confocal PL experiments when lowering the excitation level from 5×10^4 sun equivalent photon fluxes to 10^3 suns and increasing the temperature from 80 K to 300 K. In Fig. 8.10 confocally recorded PL yields for different excitation levels Φ are exemplarily outlined [17], of which some of the maxima and minima have been plotted as $\log(Y_{pl})$ vs $\log(\Phi)$. From the extrapolation across the excitation range so far not experimentally accessible toward 1 sun with expectable slopes, e.g., of the regimes where PL has been detectable, and with the "worst-case" slope of unity, the minimum regime of lateral PL variation at 1 sun excitation and 300 K is estimated, still having assumed the ratio

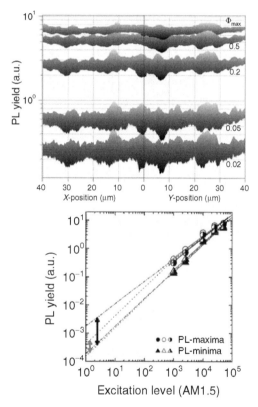

Fig. 8.10. Lateral scans $(40\mu m)^2$ of PL yields of a polycrystalline $Cu(In_{0.72}Ga_{0.28})Se_2$ (Mo/CIGS/CdS) at $T = 83$ K with variation of excitation level within $10^3(AM1.5) \leq \Phi \leq \Phi_{max} = 5 \times 10^4(AM1.5)$-equivalent photon fluxes and extrapolation to AM1.5-equivalent excitation; *arrows* indicate minimum and maximum variation of PL yield ratios $(Y_{pl,max}/Y_{pl,min})$

of nonradiative to radiative transitions to be not dependent on temperature: $kT\ln(2.7) = 26\,\mathrm{meV} \leq \Delta(\varepsilon_{\mathrm{Fn,max}} - \varepsilon_{\mathrm{Fn,min}})_{28\%\mathrm{Ga}}$ and $kT\ln(2.2) = 21\,\mathrm{meV} \leq \Delta\varepsilon_{(\mathrm{Fn,max} - \varepsilon_{\mathrm{Fn,min}})_{50\%\mathrm{Ga}}}$. When taking the temperature dependence of radiative and defect-controlled nonradiative recombination rates into account, the lateral variation $\Delta\varepsilon_{(\mathrm{Fn,max} - \varepsilon_{\mathrm{Fn,min}})}$ undoubtedly will increase (see [17]).

8.6 Analysis of Spatial Inhomogenities by Erosion–Dilatation Operations

A quantitative measure of the 2-dimensional properties of scans of AFM topology, optical reflection, and luminescence can be extracted from the recorded data by, e.g., 2-D Fourier transforms or erosion–dilatation operations in the Minkowski space (opening procedure, see Fig. 8.11), which also allow for the comparison of general structural features such as predominant spatial frequencies or frequently occurring spatial sizes.

- **Translation**: set $\underline{A}$, vector **x**

$$\underline{A} + \overline{x} = \left\{ \underline{a} + \overline{x} \middle| \underline{a} \in \underline{A} \right\}$$

- **Negation**: set $\underline{B}$, $--\underline{B} = \left\{ -\underline{b} \middle| \underline{b} \in \underline{B} \right\}$

- **Minkowski-addition-dilatation** $\hat{D}(\underline{A}, \underline{B})$

$$\hat{D}(\underline{A}, \underline{B}) = \underline{A} \oplus \underline{B} = \bigcup_{b \in B} \underline{A} + \underline{b}$$

- **Minkowski-subtraction** $\hat{D}^{*}(\underline{A}, \underline{B})$

$$\hat{D}^{*}(\underline{A}, \underline{B}) = \underline{A} \ominus \underline{B} = \bigcap_{b \in B} \underline{A} + \underline{b} = \left\{ \overline{x} \middle| -\underline{B} + \overline{x} \subset \underline{A} \right\}$$

- **Erosion** $\hat{E}(\underline{A}, \underline{B})$

$$\hat{E}(\underline{A}, \underline{B}) = \underline{A} \ominus (-\underline{B}) = \left\{ x \middle| \underline{B} + \overline{x} \subset \underline{A} \right\}$$

- **Opening** $\hat{O}(\underline{A}, \underline{B})$

$$\hat{O}(\underline{A}, \underline{B}) = [\underline{A} \ominus (-\underline{B})] \oplus \underline{B} = \hat{D}\left[\hat{E}(\underline{A}, \underline{B}), \underline{B}\right]$$

- **Closing** $\hat{C}(\underline{A}, \underline{B})$

$$\hat{C}(\underline{A}, \underline{B}) = [\underline{A} \oplus (-\underline{B})] \ominus \underline{B} = \hat{E}\left[\hat{D}(\underline{A}, \underline{B}), \underline{B}\right]$$

Fig. 8.11. Two-dimensional operations for the extraction of particular features of 2-D data sets, e.g., scan of AFM topology, optical reflection, and PL yields [18]

In Figs. 8.12 and 8.13, 2-D Fourier transforms of AFM topology scans and optical reflection scans of Cu(InGa)Se$_2$ absorbers on glass with CdS window layers are outlined for comparison. A pronounced dependence on Ga content, is also intuitively observable in Figs. 8.4, 8.14–8.16, which show considerable change in surface morphology due to Ga incorporation. So far, however, there is no experimental evidence whether the changes in morphology is related to the increase in Ga concentration by itself or initiated by the technological process of CIGS deposition.

We also detect lateral variations of the PL yield $[Y_{\mathrm{pl,max}}/Y_{\mathrm{pl,min}}]$ up to a factor of 5 for CIGS outlined in Figs. 8.14–8.16 (and even up to a factor > 10 for $x = 0.3$ [17]).

The spatial structure of the PL yield in that set of samples turns out to be much larger than the topological (AFM, reflection). Moreover, the PL scans of $(100\,\mu\mathrm{m})^2$ undoubtedly are too small to contain a data set representative of the sample properties in terms of luminescence.

The related representation in "opening functions" displayed in Fig. 8.17 shows substantial differences in PL yield patterns vs Ga content, which are represented in the respective distributions of sizes and mean diameters.

8.7 Novel Method for Depth Resolution of the Quality of the Photo Excited State

In laterally nonresolved luminescence emitted from Cu(InGa)Se$_2$ absorbers recorded with relatively large focal lengths and accordingly comparatively small solid angles, we see pronounced interference effects (Fig. 8.18). Despite the topological roughness outlined in Fig. 8.4, photons with wavelengths much larger than the size of geometrical features, according to effective medium theory, only reach interfaces sufficiently flat for the creation of interference patterns.

The spectral interference pattern has been modeled via a conventional 1-D matrix transfer formalism and the respective superposition of waves emitted from luminescence centers located at $z = z_i$ in the CIGS thin film. The photons propagate from the centers to the right-hand and left-hand sides, where they are reflected by the contrast in refractive indices at $x = 0$ (rear contact) or $x = d$ (front interface between CdS and air) (Fig. 8.19).

Since the thickness of the CdS window layer amounts only to about 50 nm, its influence on the propagation of waves with wavelengths of about 0.4–0.5 µm (0.8–1.0 eV) in CdS can be neglected. The resulting modulation function $\Phi_{\mathrm{r}}(k(\omega))$ of the emitted luminescence, including absorption in CIGS and being

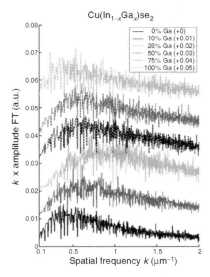

Fig. 8.12. Amplitude of 2-D Fourier transforms of AFM-topology scans $(20\,\mu m)^2$ of $Cu(In_{1-x}Ga_x)Se_2$ absorbers (amplitude multiplied by spatial wave vector k); scattering of transformed functions indicate that $(20\,\mu m^2)$ regimes might be too small to contain representative data sets

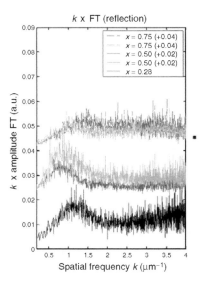

Fig. 8.13. Amplitude of 2-D Fourier transforms of scans of optical reflection $(100\,\mu m^2)$ of $Cu(In_{1-x}Ga_x)Se_2$ absorbers (amplitude multiplied by spatial wave vector k); predominant reflection features occur in the neighborhood of $k = (0.7 - 1.3)\,\mu m$; transformed functions of reflection signals recorded at different local positions show that $(100\,\mu m)^2$ regimes contain representative data sets

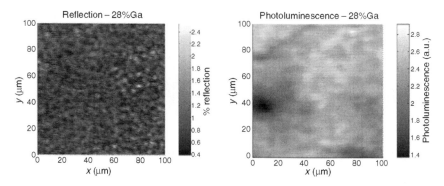

Fig. 8.14. Lateral Scans $(100\mu m)^2$ of optical reflection (*left*) and of PL (*right*) of $Cu(In_{0.72}Ga_{0.28})Se_2$ absorbers glass/CIGS $(2\,\mu m)$/CdS

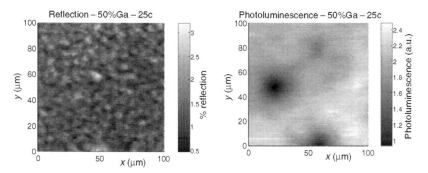

Fig. 8.15. Lateral Scans $(100\,\mu m)^2$ of optical reflection (*left*) and of PL (*right*) of $Cu(In_{0.50}Ga_{0.50})Se_2$ absorbers (glass/CIGS $(2\,\mu m)$/CdS)

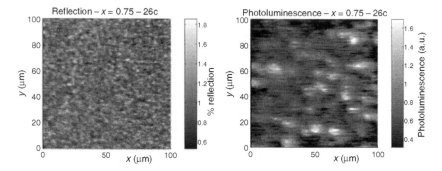

Fig. 8.16. Lateral scans $(100\,\mu m)^2$ of optical reflection (*left*) and of PL (*right*) of $Cu(In_{0.25}Ga_{0.75})Se_2$ absorbers (glass/CIGS $(2\,\mu m)$/CdS)

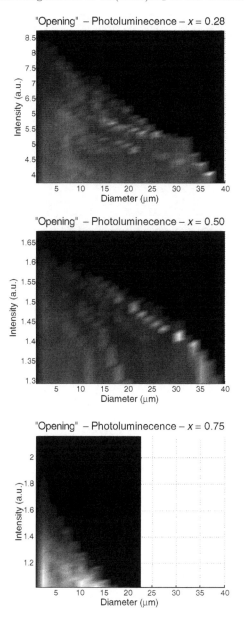

Fig. 8.17. Opening functions for the representation of features in 2-D PL yields of Cu(In$_{1-x}$Ga$_x$)Se$_2$ absorbers (glass/CIGS/CdS, see Figs. 8.14–8.16). Opening functions show on the *abscissa axes* the size of features (here of spherical shape), and on the *ordinate axes* the "height" (magnitude) of the respective signal; the gray-scale expresses the magnitude of feature sizes (fraction, how often these are available) that have changed/disappeared as the probe size gets larger

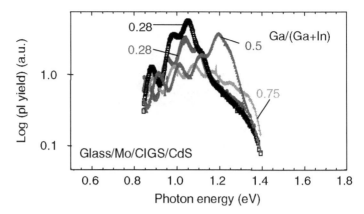

Fig. 8.18. Spectral PL yields from Cu(InGa)Se₂ absorbers on Mo rear contacts and with CdS window layer recorded without lateral resolution and by according sufficiently large focal lengths in order to provide for the conservation of phase relation between forward and backward propagating waves

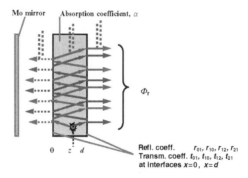

Fig. 8.19. Schematic representation of forward and backward traveling PL photons emitted from a luminous center located at z in CIGS (detection of PL spectra in terms of Φ_r towards the right-hand side)

composed of the $(i_{max} - i_{min} + 1)$ contributions of individual centers located at z_i reads as

$$\Phi_r = \left(\frac{1}{i_{max} - i_{min} + 1}\right)$$

$$\left| (t_{12})^2 \sum_{z_i=z_{i,min}}^{z_i=z_{i,max}} \exp\left[-2\alpha^* (d - z_i)\right] \left[\frac{1 + r_{10} \exp\left[-2\alpha^* z_i\right] \exp(i2k_1 z_i)}{1 + r_{10}r_{12} \exp\left[-2\alpha^* d\right] \exp(i2k_1 d_1)}\right]\right|^2,$$

$$(8.22)$$

where k, α^*, r_{mn} and t_{mn} indicate corresponding wave vectors $k(\omega)$, attenuation coefficient $\alpha^*(\omega)$, reflection and transmission factors for amplitudes in and between corresponding media, respectively.

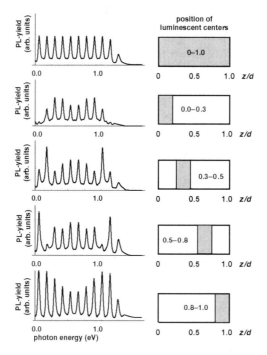

Fig. 8.20. Superposition of the contribution of luminescent centers to the spectral PL signal located at different depth positions in the CIGS absorber $\Phi_r(k(\omega))$

A preliminary fit of our simulated PL pattern to the experimental spectral luminescence results (Fig. 8.20) from the layer Mo/CIGS/CdS/air points toward an approximately homogenous profile of "luminosity" in $0.2 \leq (z/d) \leq 1.0$, which reads in approximately flat quasi-Fermi level of electrons in the CIGS absorber except close to the rear contact; further detailed studies of the drop in PL luminosity in the neighborhood of the rear contact $(x = 0)$ might indicate the magnitude of one component of the anisotropic minority carrier diffusion length in the direction of the carrier collection; this number is of crucial importance for the structure being operated as a solar cell.

8.8 Conclusions

In polycrystalline semiconductors like Cu(InGa)Se₂, spatial inhomogeneities in the μm and sub-μm scales, e.g., of crystal orientation and even of crystal composition, initiating a variation in metallurgical phases have to be expected. As a consequence of these spatial inhomogeneities, lateral variations in optoelectronic properties including those most relevant for the quantum conversion of light are very likely, such as different local bulk and interface defect densities, lifetimes and accordingly the chemical potential of the electron–hole ensemble, which in semiconductors used to be the splitting of the quasi-Fermi levels $(\varepsilon_{\mathrm{Fn}} - \varepsilon_{\mathrm{Fp}}) \geq V_{\mathrm{oc}}$.

The averaging of defect densities in almost each of the experiments for the study of defect densities and of photovoltaic properties by the geometrical boundary conditions of comparatively large spatial regimes, like contacts, may mask the local dependence of magnitudes and may even mislead the interpretation by the superposition with nonlinear interplay of the individual local effects. In particular, the spatial superposition of locally distributed open-circuit voltages V_{oc} through the equipotential surface of front and rear contact generate additional entropic terms and reduce the efficiency of the respective hetero diode solar cells.

Acknowledgments

Financial support via BMBF contract No. 01 SF0115 is gratefully acknowledged. The author wishes to thank R. Kniese (ZSW Stuttgart) for sample preparation, in particular for various variations in electronic and geometrical sample design.

References

1. Ullal, H., Zweibel, K., von Roedern, B.: Polycrystalline thin film photovoltaics: research, development and technologies. In: Benner, J. (ed.) Conference Records of the 20th IEEE Photovoltaic Specialist Conference, pp. 472–477. IEEE, Piscataway, NJ (2002)
2. Contreras, M.A., Egaas, B., Ramanathan, K., Hiltner, J., Swartzlander, A., Hasoon, F., Noufi, R.: Progress toward 20% efficiency in Cu(In,Ga)Se$_2$ polycrystalline thin-film solar cells. Prog. Photovolt.: Res. Appl. 7, 311–316 (1999)
3. Powalla, M., Voorwinden, G., Dimmler, B.: Continuous Cu(In,Ga)Se$_2$ deposition with improved process control. In: Ossenbrink, H.A. et al. (eds.) Proceedings of the 14th EU PVSEC, pp. 1270–1273. H. S. Stephens & Associates, Bedford (1997)
4. Würfel, P.: Physics of Solar Cells, Wiley-VCH, Weinheim, (2004)
5. Niles, D.W., Ramanathan, K., Granata, J., Hasoon, F., Noufi, R., Tielsch, B.J., Fulghum, J.E.: Na impurity chemistry in photovoltaic CIGS thin-films: an investigation with photo- and Auger electron spectroscopies solar cells. In: Procedings of the MRS Symposium, vol. 485, pp. 179–184. Materials Research Society, Pittsburg, PA (1998)
6. Bothe, K., Bauer, G.H., Unold, T.: Spatially resolved photoluminescence measurements on Cu(In,Ga)Se$_2$ thin films. Thin Solid Films 403–404, 453–456 (2001)
7. Bauer, G.H., Bothe, K., Unold, T.: Open circuit voltage from quasi-Fermi level splitting in polycrystalline Cu(In,Ga)Se$_2$ films by photoluminescence with lateral sub-micron resolution. In: Benner, J. (ed.) Conference Records of the 29th IEEE Photovoltaic Special Conference, pp. 700–703. IEEE, New York (2002)

8. Dhere, N.G., Ghongadi, S.R., Pandit, M.B., Kadam, A.A., Jahagirdar, A.H., Gade, V.S.: AFM, micro-PL, and PV analyses of CuIn$_{1-x}$Ga$_x$S$_2$ thin films solar cells on stainless steel foil. In: Benner, J. (ed.) Conference Records of the 29th IEEE Photovoltaic Special Conference. IEEE, New York (2002)
9. Powalla, M., Dimmler, B.: Scaling up issues of CIGS solar cells. Thin Solid Films 361–362, 540–546 (2000)
10. Powalla, M., Dimmler, B.: CIGS solar cells on the way to mass production: process statistics of a 30 cm × 30 cm module line. Sol. Energy Mater. Sol Cells 67, 337–344 (2001)
11. Voorwinden, G., Kniese, R., Powalla, M.: In-line Cu(In,Ga)Se$_2$ co-evaporation processes with graded band gaps on large substrates. Thin Solid Films 431–432, 538–542 (2003)
12. Kniese, R., Hariskos, D., Vorwinden, G., Rau, U., Powalla, M.: High band gap Cu(In,Ga)Se$_2$ solar cells and modules prepared with in-line co-evaporation. Thin Solid Films 431–432, 543–547 (2003)
13. van Roosbroeck, W., Shockley, W.: Photon-radiative recombination of electrons and holes in germanium. Phys. Rev. 94, 1558–1560 (1954)
14. Schick, K., Daub, E., Finkbeiner, S., Würfel, P.: Verification of a generalized Planck law for luminescence radiation from silicon solar cells. Appl. Phys. A 54, 109–114 (1992)
15. Smestad, G., Ries, H.: Luminescence and current–voltage characteristics of solar cells and optoelectronic devices. Sol Energy Mater. Sol Cells 25, 51–71 (1992)
16. Unold, T., Berkhahn, D., Dimmler, B., Bauer, G.H.: Open circuit voltage and loss mechanisms in polycrystalline Cu(In,Ga)Se$_2$-heteridiodes from photoluminescence studies. In: Scheer, H. et al. (eds.) Proceedings of the 16 EU PVSEC, pp. 736–739. James & James Science Publications, London (2000)
17. Bauer, G.H., Gütay, L., Kniese, R.: Structural properties and quality of the photoexcited state in Cu(In$_{1-x}$Ga$_x$)Se$_2$ solar cell absorbers with lateral submicron resolution. Thin Solid Films 480, 259–263 (2005)
18. Dudgeon, D.E., Mersereau, R.M.: Multidimensional Digital Signal Processing. Prentice-Hall, Englewood Cliffs, New Jersey (1984)
19. Giardina, C.R., Dougherty, E.R.: Morphological Methods in Image and Signal Processing. Prentice-Hall, Englewood Cliffs, New Jersey (1988)
20. Heijmans, H.J.A.M.: Morphological Image Operators. Advances in Electronics and Electron Physics. Academic Press, Boston (1994)
21. Serra, J.: Image Analysis and Mathematical Morphology. Academic Press, London (1982)

9

Electro-Optical Properties of the Microstructure in Chalcopyrite Thin Films

N. Ott, H.P. Strunk, M. Albrecht, G. Hanna, and R. Kniese

We show the usefulness of transmission electron microscopy for the investigation of the microstructure and its electro-optical properties of various Cu(In,Ga)Se$_2$ (CIGS) thin films on a microscopic scale. The films have different grain orientations (texture) and Na contents. The electro-optical properties are analyzed with a cathodoluminescence (CL) spectrometer that is attached to the transmission electron microscope (TEM). We find that the $\{2\,2\,0/2\,0\,4\}$-oriented films show almost no activity of the grain boundaries and should thus be well suited for solar cell applications. First results are also presented on the irradiation damage that occurs due to the electron bombardment of the films. These results open a novel approach to study in situ the generation of atomic defects in an electron microscope.

9.1 Introduction

There is an intimate relationship in semiconductor materials between microstructure and electronic properties. The TEM is an ideal, and thus in many cases common, instrument to analyse the microstructure. Conventional TEM imaging permits obtaining a detailed view on the grain structure, the dislocations and stacking faults. Only atomic defects escape such an analysis, unless they occur in agglomerates. We have extended the potential of our analytical TEM by attaching to it a full CL spectrometer, which delivers information on the local optoelectronic properties of the sample, in addition to the structural information (and chemical information obtained by energy dispersive X-ray analysis and parallel electron energy loss spectrometry). As measuring the luminescence in a TEM is a rather new technique, we will first briefly describe here the experimental setup. We will then exemplify the rewarding potential of this technique by presenting first results obtained on Cu(In,Ga)Se$_2$ absorber films, used in thin-film solar cells. In addition to the analysis of the optoelectronic properties of the extended defects, we will also show that CL–TEM opens a new way to study in situ electron irradiation damage.

9.2 Experimental Procedures

9.2.1 Sample Preparation

Material analysis in a TEM requires electron-transparent samples that need to be specially prepared to a final thickness of between ten and a few hundred nanometers, depending on the envisaged analysis. The investigations reported in this chapter utilize plan view samples prepared from CIGS films deposited on Mo-covered glass. These samples, which are viewed in the direction perpendicular to the CIGS film, are prepared by first cutting from the film, square pieces approximately 3 mm across. These pieces are then ground from the rear side with diamond foils of decreasing graining until all of the glass is removed, i.e., only the CIGS film with underlying Mo layer is left. A nickel ring is then glued to the sample to increase the mechanical stability. The preparation is finished by Ar-ion milling this sample under liquid nitrogen cooling until a hole appears in the center. Its rim is usually thin enough for analysis by TEM.

9.2.2 The Transmission Electron Microscope

We use for our work a Philips CM30 TEM with a maximum accelerating voltage of 300 kV and a spatial (point) resolution of 0.23 nm. The microscope is equipped with a scanning unit, an energy dispersive X-ray spectrometer and a parallel electron energy loss spectrometer for simultaneous acquisition of structural and chemical data. For the CL analyses an Oxford MonoCL2 system is attached to this microscope. The light emitted from the sample is collected by a parabolic half-mirror that is inserted into the gap of the objective pole piece above the specimen (see sketch in Fig. 9.1). An evacuated tube guides the light through a window into the spectrometer consisting of two interchangeable gratings and two detectors.

The use of an optical chopper for lock-in applications is provided. The collected light can also directly be measured in the whole spectral range for panchromatic analysis. The two detectors are a photomultiplier tube operating in the visible and ultraviolet part and a Ge detector for the infrared part of the spectrum. The system is thus capable to detect light within the wavelength range of 180–1,800 nm (energy range 6–0.6 eV). As the temperature is important in CL analysis, we have two cooling goniometer holders available (liquid nitrogen or liquid helium cooled) that permit to adjust the specimen temperature within the range of 10–300 K. In general, we have used a temperature of 90 K and an accelerating voltage of 150 kV for the CL work. The bright field micrographs of the CL-investigated areas have been taken at 300 kV in imaging mode.

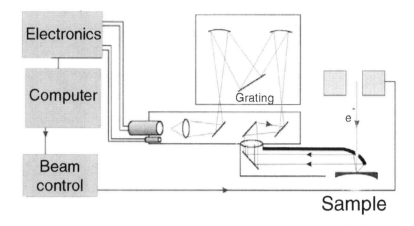

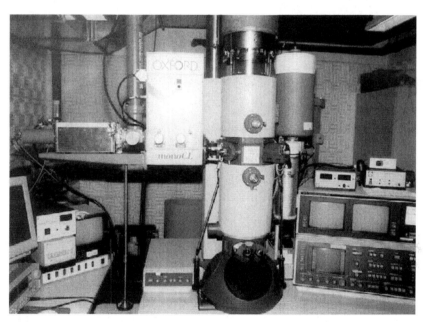

Fig. 9.1. Sketch of the cathodoluminescence unit (*top*). The illuminating electron beam e⁻ is focused by the objective lens onto the sample through a small hole in the parabolic mirror. The emitted light leaves the microscope column through an evacuated pipe and enters the spectrometer by a lens. The *bottom picture* shows the unit as attached to the transmission electron microscope at the *left* to the objective section of the column. An X-ray detector system is visible at the *right*

9.2.3 Cathodoluminescence Analysis in the Transmission Electron Microscope

In view of the relatively new experimental approach, we would like to consider a few aspects of CL in the TEM in some detail in this chapter. Our instrument permits two different operating techniques – the spectrum acquisition from a certain area or the mapping of the sample in the light of a specific wavelength interval. For the spectrum acquisition, the area of interest is illuminated by the electron beam and the emitted luminescence light is spectrally analyzed. The area of interest can be adjusted between a comparably large area, say a few tens of µm in diameter, down to a few nm, i.e., well below the size of a single grain. For the mapping mode, we use the scanning unit of the microscope and record the luminescence signal as the exciting electron beam scans over the selected area on the sample. In one extreme, all of the emitted light (panchromatic mode), in the other extreme, light of a specific wavelength, more precisely of a rather narrow wavelength window (monochromatic mode), is used to form an image of the intensity distribution, gray scale or by color coding. Bright field STEM images (BF) of exactly the same area can be acquired at the same time and used to compare the CL characteristics with structural properties.

This possible correlation poses the question as to the achievable lateral resolution in a micrograph produced by luminescence mapping. There is no general answer, since the lateral resolution depends not only on the diameter of the exciting electron beam, but also on the beam spread inside the sample, thus on the scattering power of the investigated material and on the specimen thickness. Beam spreads that have been calculated for four different materials using an earlier version of the Lehigh program for Monte Carlo simulation (see [1]) are displayed in Fig. 9.2 and demonstrate the characteristic dependencies on specimen thickness and average atomic weight.

Since the interesting luminescence signal follows from recombination of injected carriers, the lateral resolution depends also on the minority carrier diffusion length and – in the case of structural defects – on the recombination strength. In case the recombination strength is appreciable or even strong, the contrast width is essentially governed by the diameter of the exciting electron beam. Once the sample thickness is in the order of or smaller than the diffusion length, the two sample surfaces act as sinks and cause an effective diffusion length, depending on the surface recombination velocity, in the order of the specimen thickness. This effect can be used to improve resolution in very thin samples provided the luminescence signal is then still intense enough to be monitored. For example, in Fig. 9.4a grain boundaries are visible with a contrast width of around 60 nm, whereas grain boundaries in Fig. 9.4e exhibit contrast widths of almost 200 nm. Without going into a detailed discussion of the contrast mechanisms acting here, which appears premature in the present state of the analysis, one can conclude that obviously the respective diffusion length in the material of Fig. 9.4a is much less than in that of Fig. 9.4e.

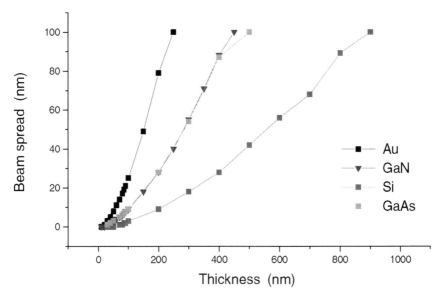

Fig. 9.2. Calculated spreading of the electron beam in dependence of specimen thickness. This example is calculated for 150 keV electron beams. The curve for GaAs holds practically for Cu(In,Ga)Se$_2$ also. At a sample thickness of say 200 nm a beam diameter of 30 nm governs the lateral resolution. For comparison in a scanning electron microscope the beam diameter (diameter of the excitation volume) is in the order of 1 μm

There is a further aspect that has to be kept in mind: CLspectra are not necessarily equivalent to photoluminescence spectra. First, the use of electrons creates many hot carriers, the recombination of which can lead to additional luminescence signals. Second, in contrast to photoluminescence experiments, the CL is generated deep in the volume of the sample. Therefore, one should always consider self-absorption of the created light inside the sample itself, which reduces the measured intensities near the band edge energy and beyond drastically. For a discussion, see [2]. In addition, the use of thin specimens in our experiments can cause Perot–Fabry oscillations due to internal reflections.

At this point, we should mention that all these influences on the contrast prevent presently the quantitative comparison of intensities and contrasts between different samples. Contrast properties can be compared between different sites within a single sample, although a way for determining the absolute scale of the intensities is still to be developed.

9.3 Results

In this chapter, we will show first results of our investigation of CIGS thin films by CL–TEM. We will start with results of films having different preferred grain orientation and Na content and then report on observations of electron radiation damage.

9.3.1 Role of Texture and Na Content

We present results on microstructure and locally resolved CL that we obtained from films with different textures and with various Na contents from different Na sources. The films have been grown by co-evaporation according to the three-stage process [3] as part of a series intended to investigate the influence of texture on solar cell efficiencies and grain boundary activity [4]. The texture of the films was determined by X-ray-diffraction. It can be controlled during the growth of the films by varying temperature and/or Se rate. For more details on growing the differently textured films, see [5]. In this chapter, we will focus on four different samples labeled: Na-free, NaF-precursor, Glass-112, and Glass-220. All samples have a Ga/(In + Ga) ratio of 0.28, are Cu poor and are all grown onto a Mo-coated glass substrate. The Na-free and NaF-precursor films contain an additional Na-barrier layer between the Mo layer and the glass. Thus, the outdiffusion of Na from the glass into both growing CIGS films is suppressed. As a consequence, the Na-free layer is, in fact, practically free of Na. On the other hand, the NaF-precursor sample has been intentionally doped with Na by adding a NaF-precursor layer on top of the Mo layer.

Figure 9.3a–d shows BF-TEM images of the different samples. All samples show a similar microstructure: the grains that can be identified from their irregular boundaries have similar size. Due to the fact that the grains frequently overlap in the sample the boundaries are sometimes difficult to trace. Also, the defect concentration inside the grains is rather similar; easily recognizable are straight structures that represent most probably twins and twin boundaries. Dislocations can rarely be identified in these contrast-rich micrographs. Contrary to these similar microstructural features, the cells made from these films exhibit different electrical properties [5] (see the efficiency values in Table 9.1).

The CL behavior of these samples yields deeper insight. Figure 9.4 shows panchromatic CL mappings and the corresponding BF-TEM images of the samples. Whereas the BF-TEM images show almost no differences between the samples as mentioned above, the CL mappings reveal great differences. The CL mapping of the Na-free sample (Fig. 9.4a) shows strong contrasts brought about by bright areas (strong luminescence) separated by dark lines (weak luminescence). When we compare the CL mapping with the BF–TEM image of the same area (Fig. 9.4b), we can identify the dark lines with the grain boundaries and other structural defects. The structural defects show decreased luminescence, thus strongly increased nonradiative recombination (as compared to the matrix).

The NaF-precursor sample shows a similar CL behavior (Fig. 9.4c). We also see bright areas separated by dark lines that correlate with grain bound-

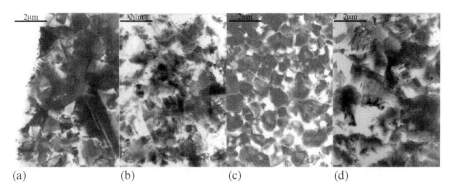

(a)   (b)   (c)   (d)

Fig. 9.3. Structure of different CIGS thin films (BF-TEM): (**a**) Na-free sample, containing no Na {2 2 0/2 0 4}-texture; (**b**) NaF precursor sample, Na from NaF precursor {1 1 2}-texture; (**c**) Glass-112, Na from glass substrate {1 1 2}-texture; (**d**) Glass-220, Na from glass substrate, {2 2 0/2 0 4}-texture. All images show a similar structure with grains around 1 μm in size containing some structural defects

Table 9.1. Summary of the results on differently prepared CIGS thin films showing different Na content, texture and thus luminescence behavior

sample	Na source	texture	CL contrast	Eff. [4]
Na free	–	{2 2 0/2 0 4}	dark at grain boundaries	5.7%
NaF	precursor	{1 1 2}	dark at grain boundaries	12.2%
Glass-112	from glass	{1 1 2}	dark at grain boundaries	
Glass-220	from glass	{2 2 0/2 0 4}	practically none	12.8%

aries and structural defects. The contrast appears somewhat blurred as compared to that of the Na-free sample probably due to a larger overlap of the grains. The Glass-112 sample shows a CL behavior (Fig. 9.4e) comparable to the NaF-precursor sample. Due to the slightly larger grains that practically do not overlap, in this case grain boundaries show up very clearly as dark lines that separate bright areas of strong luminescence.

The Glass-220 sample shows a rather different CL contrast behavior (Fig. 9.4g). The CL intensity was rather low and the contrast amplification settings had to be selected much higher as compared to the previous samples in order to detect any contrast. In this figure, it is difficult to trace the grain boundaries (compare to the BF–TEM image of the same area in Fig. 9.4h). This fact can be interpreted by a rather low nonradiative recombination activity of the grain boundaries in this specific material. In turn, a high diffusion length can be deduced from this reduced recombination activity of the grain boundaries. The low overall intensity in this sample can be attributed to prominent nonradiative recombination at the surfaces of the sample, because of this higher diffusion length.

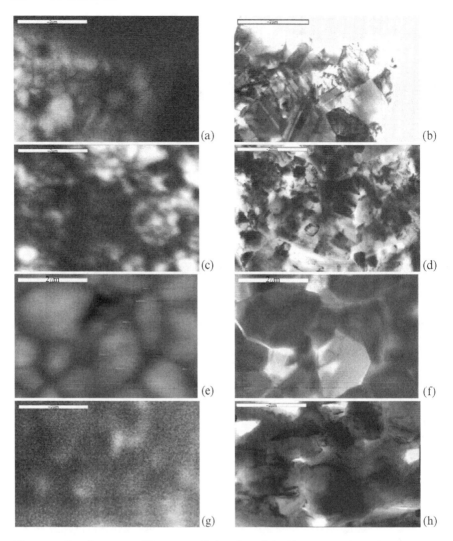

Fig. 9.4. Panchromatic CL images (*left column*) and corresponding BF–TEM images (*right column*) from the different samples: Na free (**a**, **b**), NaF precursor (**c**, **d**), Glass-112 (**e**, **f**), Glass-220 (**g**, **h**). All but the Glass-220 samples show decreased luminescence (*dark line*) at grain boundaries and other structural defects. For the Glass-220 sample no correlation is observed between CL properties and structural details

Table 9.1 summarises the results of the four samples. We see that all but the Glass-220 sample show increased nonradiative recombination (as compared to the crystalline matrix) at the grain boundaries indicated by the strong CL contrast at their positions. Thus, we consider the grain boundaries to be electrically active (and to contribute to the reduced efficiency of the cells fabricated from these materials). The Na content and the texture of the film influence the properties of the grain boundaries. The {220/204}-textured Na-containing film shows practically no contrast at the grain boundaries. At the present state of investigations we can conclude that Na can passivate the grain boundaries (see also [6]). Further clarification requires spectrally resolved analysis. The {220/204}-texture of the film, thus seems to have a particular grain boundary population that is superior to that of the {112}-textured films (see also [7]). The grain boundaries are practically inactive making this type of film the best absorber layer for cell production. This conclusion is supported by the recent record efficiency cell made by Contreras et al. [8], who used Na-containing glass and {220/204}-textured absorbers.

9.3.2 Electron Irradiation Damage

Using TEM, one always has to keep in mind that the electron beam can create irradiation damage. This damage initially occurs on an atomic scale and therefore remains invisible in the micrographs until the defect density is high enough to form agglomerates that in fact are visible in structural images. The CL–TEM system now offers the unique possibility to investigate the early stage of irradiation damage, at least in materials that luminesce, like ceramics or semiconductors, provided the created atomic defects are luminescence active. In principle, one has to discriminate two fundamentally different damage mechanisms, atomic displacement and radiolytic damage. In the first mechanism, an electron transfer has enough momentum in a knock-on interaction to an atom to displace it from a crystal lattice site into an interstitial site. There is a minimum momentum or electron energy beyond which this mechanism can occur. The damage rate increases with electron energy. In the radiolytic mechanism, the electron during passage through the specimen transfers energy into the electronic system by Coulomb interaction. This energy then subsequently causes the displacement of an atom or ion. Well-known examples are the color centers in ionic crystals (see [9]). The cross-section for Coulomb interaction falls with increasing electron velocity or energy and so does the damage rate.

Figure 9.5 shows CL spectra of CIGS samples with different Ga contents. The sample with a Ga content of 10% shows two prominent peaks, one close to the nominal band edge and the other at approximately 0.86 eV. The sample with 28% of Ga shows also two peaks but the near-band-edge peak is weakly pronounced. The other peak lies at approximately 0.97 eV. The Ga-free sample shows only one broad band centered at approximately 0.86 eV.

Figure 9.6 shows the time dependence of the CL intensity at 0.86 eV for different samples. The intensity increases with irradiation time, i.e., the electron beam generates a luminescent defect. The damaging rate decreases with increasing Ga content, at least in this low Ga-content region.

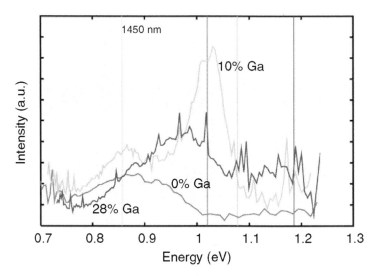

Fig. 9.5. CL spectra for samples grown with different Ga content. *Vertical lines* mark the bandgap energies, calculated from the Ga-content according to [10]

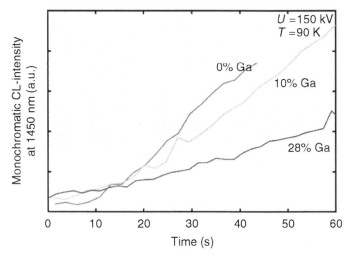

Fig. 9.6. CL intensity of the defect-related peak at 0.86 eV (1,450 nm) in dependence of irradiation time for samples with different Ga content

In order to determine the mechanism for the radiation damage, we investigated the damaging rate in dependence of the accelerating voltage of the electron beam. Figure 9.7 shows the monochromatic CL intensity at 0.86 eV of the sample with a Ga content of 10% in dependence of irradiation time and accelerating voltage (the beam current density was kept constant).

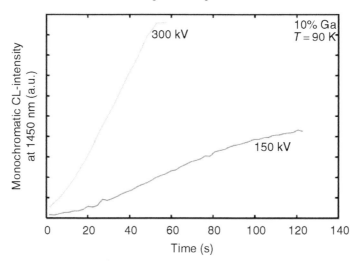

Fig. 9.7. Monochromatic CL intensity of the defect-related peak at 0.86 eV (1,450 nm) in dependence of irradiation time for two different accelerating voltages. The electron beam current density and the illuminated area have been the same at the two accelerating voltages

The radiation damage rate is higher at higher accelerating voltages. The damage is thus, as discussed above for the two types of damage mechanisms, dominated by knock-on damage. Certainly, these results are not sufficient to rule out a small contribution of radiolytic damage.

9.4 Summary and Outlook

We have shown that, depending on the deposition parameters, the grain boundaries in chalcopyrite thin films can exhibit a large variety of optoelectronic properties. These properties are essentially controlled by the defect structure that can provide recombination centers for minority carriers. The presented results indicate that careful cultivation of special textures can lead to grain boundary populations that exhibit a comparatively low nonradiative recombination rate, which is a prerequisite for improved efficiencies of solar cells. This aspect holds especially in the case of chalcopyrites where the grain size is in the order of the minority carrier diffusion length. The observations as discussed here, are first results and display the potential of our rather new analytical addition of CL to an analytical scanning TEM. As a consequence, the respective results point to a systematic analysis of optoelectronic properties of the crystal defects in chalcopyrites, which becomes feasible now with CL–TEM. The (almost) classical approach that proved very helpful in multicrystalline silicon by measuring the recombinative action of defects with electron beam induced current in a scanning electron microscope and subsequent analysis of the grain boundary types in a TEM [11] is practically not possible with chalcopyrite films because of their small grain size.

The second part of this chapter discussed first observations that indicate a novel path to the analysis of the atomic defects that are created by electron irradiation by utilizing their luminescent properties. This analysis is unique as it offers to trace in situ defect production and properties before they become visible in the electron microscope due to agglomeration. Further investigations of this type will, for example, also support the analysis and simulation of radiation properties of chalcopyrites thin-film solar cells in space.

Acknowledgments

The beam spread data in Fig. 9.2 were provided by C. Grazzi. Part of this work was carried out in the Central Facility for High Resolution Electron Microscopy of the Friedrich-Alexander-University Erlangen-Nürnberg, Germany. Financial support was by the German Federal Ministry of Education and Research (BMBF) under contract no. 01SF0113 within the framework Hochspannungsnetz (high-voltage network), a joint federal program of a number of research groups, is gratefully acknowledged. The Bavarian Ministry for Science, Research and Arts made possible the acquisition and installation of the cathodoluminescence attachment by a grant within its long-term programme "New Materials."

References

1. http://web.utk.edu/~srcutk/htm/simulati.htm
2. Holt, D.B., Yacobi, B.G.: In: Holt, D.B., Joy, D.C. (eds.) SEM Microcharacterization of Semiconductors, Vol. 12 of series "Techniques in Physics", p. 353. Academic Press, Boston, San Diego, New York, (1989)
3. Gabor, A.M., Tuttle, J.R., Albin, D.S., Contreras, M.A., Noufi, R., Hermann, A.M.: High-efficiency $CuIn_xGa_{1-x}Se_2$ solar cells made from $(In_x,Ga_{1-x})_2Se_3$ precursor films. Appl. Phys. Lett. 65, 198 (1994)
4. Hanna, G., Glatzel, Th., Sadewasser, S., Ott, N., Strunk, H.P., Rau, U., Werner, J.H.: Texture and electronic activity of grain boundaries in polycrystalline Cu(In,Ga)Se$_2$ thin films. Appl. Phys. A (in print), published online: DOI: 10.1007/s00339-005-3411-1
5. Hanna, G., Mattheis, J., Laptev, V., Yamamoto, Y., Rau, U., Schock, H.W.: Influence of the selenium flux on the growth of Cu(In,Ga)Se$_2$ thin films. Thin Solid Films 431–432, 31 (2003)
6. Ott, N., Hanna, G., Albrecht, M., Rau, U., Werner, J.H., Strunk, H.P.: Cathodoluminescence studies of Cu(In,Ga)Se$_2$ thin-films. In: Sameshima, T., Fuyuki, T., Werner, J.H., Strunk, H.P. (eds.) Polycrystalline Semiconductors VII. Solid State Phenom. 93, 133 (2003)

7. Ott, N., Hanna, G., Rau, U., Werner, J.H., Strunk, H.P.: Texture of Cu(In,Ga)Se$_2$ thin films and nanoscale cathodoluminescence. J. Phys.: Condens. Matter 16, 85 (2004)
8. Contreras, M.A., Egaas, B., Ramanathan, K., Hiltner, J., Swartzlander, A., Hasoon, F., Noufi, R.: Progress towards 20% efficiency in Cu(In,Ga)Se$_2$ polycrystalline thin-film solar cells. Progr. Photovolt.: Res. Appl. 7, 311 (1999)
9. Crawford, J.H. Jr.: Radiolysis of alkali halides. Adv. Phys. 17, 93 (1968)
10. Alonso, M.J., Garriga, M., Rincon, C.A.D., Hernandez, E., Leon, M.: Optical functions of chalcopyrite CuGa$_x$In$_{1-x}$Se$_2$ alloys. Appl. Phys. A 74, 659 (2002)
11. Cunningham, B., Strunk, H., Ast, D.G.: First and second order twin boundaries in EFG silicon ribbon. Appl. Phys. Lett. 40, 237 (1982)

10

Electronic Properties of Surfaces and Interfaces in Widegap Chalcopyrites

S. Sadewasser

Chalcopyrite solar cells consist of a layered stack of several metallic and semiconducting materials. As light-induced charge carriers are transported perpendicular to the layers, the interfaces between the layers are of paramount importance for device efficiency. This chapter describes the application of Kelvin probe force microscopy to research on chalcopyrite materials and solar cell devices. The technique is used to measure the sample's work function with a lateral resolution in the nm range. After introducing the experimental technique, we review results on surface inhomogeneities on the local scale, including orientation-dependent work function, foreign phases and grain boundaries. In a separate section measurements on complete solar cell devices in a cross-sectional arrangement are introduced and discussed.

10.1 Introduction

Thin-film solar cells based on Cu-chalcopyrite absorber material presently achieve efficiencies of more than 19% on the laboratory scale [1], with several efforts being made to build up pilot production lines. The absorber material used mainly consists of the compound $Cu(In,Ga)(S,Se)_2$ or the sulfur-free $Cu(In,Ga)Se_2$, usually showing an energy gap in the range of 1.1 eV [2, 3]. Cells with larger bandgap are required for the top cell in tandem devices. However, the open-circuit voltage V_{oc} does not increase proportionately to the bandgap, leading to losses in the power conversion efficiency of those cells [4]. As the structure of chalcopyrite solar cells is rather complicated, the reasons for the deviation between bandgap and V_{oc} are not known yet and are under current investigation. A typical chalcopyrite solar cell consists of a substrate (glass), a back contact (Mo), a p-type absorber material, a thin buffer layer (CdS), an n-type window layer (ZnO) and a front contact (NiAl). Whereas the back contact, window layer and front contact are deposited by well-known and well-controlled processes (sputtering or evaporation), the deposition of the absorber layer is achieved by a variety of different processes, including sequential and/or coevaporation, sequential sputtering and subsequent rapid thermal annealing and a variety of chemical processes such as (metal organic) chemical vapor deposition (CVD), electrodeposition, etc. The buffer layer is also under current investigation, exploring different materials (ZnSe, ZnS, CdS) deposited by various processes (chemical bath deposition, etc.). As growth processes vary widely, so do the sources for possible losses in the device. One source for losses is the interfaces between the various layers, the interface between absorber and buffer layer being especially sensitive. Therefore, with the goal of understanding the specifics of the surfaces and interfaces, a study of their properties by surface sensitive techniques is crucial. An overview of the surface and interface properties has been given by Scheer [5], including a discussion of surface orientation, surface composition and surface states, as well as the effects of adsorption, chemisorption and chemical etching. An overview of several studies on the band line-up between absorber and buffer is also given, reviewing studies employing photoemission and surface photovoltage (SPV). However, the absorber material is a polycrystalline thin film, with grain sizes of the order of $1\,\mu$m. Thus, studies on this length scale are required to develop a detailed understanding of the material properties.

Atomic force microscopy (AFM) provides access to such a small lateral scale; it was developed through the last 20 years and provides now a standard characterization technique for surface structure in many laboratories [6, 7]. Besides several other specialized AFM variations, the Kelvin probe force microscopy (KPFM) is designed to obtain also electronic surface information. By a compensation of the electrostatic forces, the contact potential (CP) of the surface is measured in addition to the regular topographic characterization [8, 9]. KPFM has been widely applied in recent years for semiconductor characterization, ranging from Si device cross-sections [10] to III–V surface studies [11] and organic semiconductors [12]. From CP measurement it is in principle possible to deduce the work function of the material by calibrating the AFM tip against a known surface. However, since the electronic surface properties are extremely sensitive to the condition, i.e., the cleanness of the surface, it is advisable to work in ultrahigh vacuum (UHV) in order to obtain meaningful work function measurements. Besides the advantage of studying clean surfaces, KPFM in UHV also offers an enhanced lateral resolution (down to 20 nm), mainly due to a smaller distance between tip and sample [13].

In this chapter the application of the KPFM method to the study of chalcopyrite solar cell materials will be presented. Initially, the KPFM method itself will be introduced. Then, an overview of the various KPFM studies performed on chalcopyrites will be given; the possibility for cleaning surfaces will be discussed, followed by studies on surface orientation and different surface inhomogeneities. The presentation of experimental results concludes with a section about cross-sectional studies on complete solar cell devices. Outlook will summarize the results and discuss possible future applications of the KPFM method to chalcopyrite solar cell research.

10.2 Experimental Method

Noncontact AFM (NC-AFM) measures the topography of the samples by means of the change of the cantilever's resonance frequency when tip–sample forces act on the cantilever [7]. In order to maintain the tip–sample distance constant, two distinct modes can be used. In the amplitude modulation technique (AM-mode) the tip–sample distance is changed so that the amplitude of the oscillation is maintained constant. For the frequency modulation technique (FM-mode) the frequency shift with respect to the free resonance frequency is maintained constant by adjusting the tip–sample distance [14]. The fact that the cantilever is sensitive to the force gradient is responsible for the high resolution of AFM [6]. Due to the higher quality factor of the oscillation and the related rapid decrease of the oscillation amplitude, the FM-mode has to be used in UHV [14].

KPFM is based on NC-AFM and measurement of the work function employs the electrostatic forces between tip and sample [8, 9]. A voltage is applied between tip and sample, consisting of a DC-bias V_{DC} and an AC-voltage $V_{AC} \sin(\omega t)$ at the frequency ω. The application of this voltage, $V(t)$, results in an oscillating electrostatic force, inducing an additional oscillation of the cantilever at ω. Assuming the tip–sample system to be a capacitor, the electrostatic force can be written as:

$$F_{es} = -\frac{1}{2}\frac{\partial C}{\partial z}V(t)^2 = -\frac{1}{2}\frac{\partial C}{\partial z}\left[V_{DC} - V_{CP} + V_{AC}\sin(\omega t)\right]^2, \tag{10.1}$$

where $\partial C/\partial z$ is the gradient of the capacitance of the tip–sample system and $V_{CP} = \Delta\Phi/e$ the contact potential, where e is the elementary charge, which is the difference in work function between tip and sample. Equation 10.1 can be written as $F_{es} = F_{DC} + F_\omega + F_{2\omega}$, with the spectral components:

$$F_{dc} = -\frac{\partial C}{\partial z}\left[\frac{1}{2}\left(V_{DC} - V_{CP}\right)^2 + \frac{V_{AC}^2}{4}\right], \tag{10.2}$$

$$F_\omega = -\frac{\partial C}{\partial z}\left(V_{DC} - V_{CP}\right)V_{AC}\sin(\omega t), \tag{10.3}$$

$$F_{2\omega} = \frac{\partial C}{\partial z}\frac{V_{AC}^2}{4}\cos(2\omega t). \tag{10.4}$$

The DC part F_{DC} in Eq. 10.2 contributes to the topographical signal, the term F_ω at the AC frequency ω in Eq. 10.3 is used to measure CP and the contribution $F_{2\omega}$ in Eq. 10.4 can be used for capacitance microscopy. A lock-in amplifier is used to detect the cantilever oscillation at ω. This signal is minimized by applying the DC bias V_{DC}, which corresponds to CP as can be seen from Eq. 10.3; recording V_{DC} with the scan gives an image of CP. Calibrating the tip against a reference (e.g., highly oriented pyrolytic graphite) with a known work function allows derivation of the sample's work function.

The present setup is a modified Omicron UHV-AFM/STM operating at a base pressure $<10^{-10}$ mbar [15,16]. For CP measurement AM-mode is used, which detects the long-range electrostatic force directly. In our experimental setup, the AC frequency ω is tuned to the second resonance frequency of the cantilever, which results in improved sensitivity and allows the independent and simultaneous imaging of the topography and CP. An energy resolution of $\sim$5 meV is obtained with AC voltages as low as 100 mV. This allows application to semiconductors, where large AC voltages could induce band bending at the surface, thereby falsifying the work function measurement.

Additional information about surface band bending can be obtained by the SPV method [17]. The sample is illuminated by a laser, resulting in charge carrier excitation if the laser energy is above the bandgap. We used a red laser with $\lambda = 675$ nm (20 mW max. intensity) or a UV laser with $\lambda = 344$ nm (70 mW max. intensity). As the occupation of surface states changes, the surface band bending is changed and in most cases reduced. If the illumination intensity is sufficient, flat band conditions at the surface can result [17].

10.3 Results and Discussion

10.3.1 Surface Condition and Cleanness

Prior to the investigation of specific surface properties, a concise assessment of the surface condition and the surface cleanness should be performed. Best conditions would be obtained for growth and transfer of the prepared sample surface under UHV conditions (see below), although this is not possible in most cases. Most of the growth systems from which samples were studied do not allow a direct transfer into the UHV system. Thus, exposure of the sample surface to air is unavoidable in many studies. In some examples (see below), the samples were transferred in inert gas, thereby avoiding water condensation and chemical reactions of the surface with oxygen from the air.

For the cases where samples had to be transferred through air, a cleaning procedure was developed, comprising annealing and sputtering treatments. Exemplarily, a cycle of several subsequent annealing and sputtering treatments on a CuGaSe$_2$ surface will be presented here. A $\sim$1.5 μm-thick CuGaSe$_2$ thin film was grown on Mo-coated soda lime glass by physical vapor deposition (PVD) [18]. The sample had to be exposed to air before introduction into the UHV, in order to mount it on the sample holder. In Fig. 10.1a the topography image as measured with the KPFM is presented. Various grains with grain sizes between 200 and 700 nm can be seen. Height differences up to 360 nm within the image indicate that the substrate is completely covered and the surface is fairly flat (when compared to the grain height of 1.5 μm, as concluded from cross-sectional scanning electron microscopy images, not shown). The simultaneously measured work function image for dark conditions is shown in Fig. 10.1b.

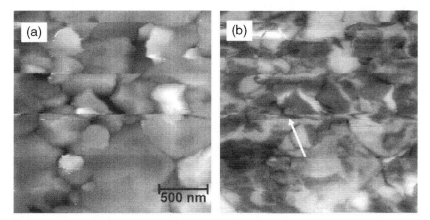

Fig. 10.1. KPFM measurement on PVD-grown CuGaSe$_2$: (**a**) topography shows the grains of the polycrystalline thin film (gray scale = 360 nm); (**b**) the simultaneously measured work function ($\Phi = 4.23 - 4.39$ eV). The *white arrow* gives the position of the linescan in Fig. 10.7 Reprinted from [19] with permission

Subsequent to this initial measurement the sample was cleaned by first annealing it at 170°C for 1 h, then sputtering the surface with Ar ions at 500 keV for 90 s (incident angle varied between 45° and 90° to the surface), again annealing (180°C for 30 min), another sputtering step (500 eV for 10 min) and a final annealing step at 180°C for 30 min. After each treatment the sample surface was imaged by KPFM and the work function determined. Measurements were performed in the dark and under illumination (20 mW at $\lambda = 675$ nm). An overview of the averaged work function obtained from the images is presented in Fig. 10.2.

The initial low work function of ∼4.31 eV is increased to ∼5.00 eV after removal of the contamination water layer by the first annealing treatment. Subsequent sputtering lowers the work function slightly. This is explained by the creation of defects due to the Ar-ion impact on the surface. This sample, however, shows a higher SPV, reaching nearly the same work function under illumination as the annealed sample. Under illumination charges at the surface, defects are neutralized by photogenerated charge carriers. The band bending is reduced and the work function increases. The subsequent annealing treatment increases the work function in the dark; however, only very small SPV is observed. Due to annealing the atoms at the surface become more mobile, which results in defect healing. This reduces the band bending and accordingly also the SPV. Subsequent sputtering and annealing steps only change the work function slightly, showing again the sputter-induced defects and a high SPV. From these results it can be concluded that (1) it is very important to clean the sample's surface prior to measurement of the work function and (2) a combination of annealing and sputtering results in a saturated and stable surface condition.

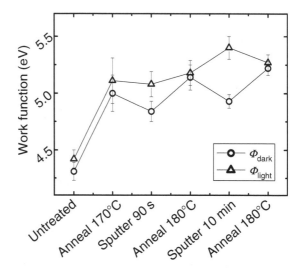

Fig. 10.2. Work function evolution after subsequent annealing and Ar-ion sputtering treatments of the surface of a PVD-grown CuGaSe₂ absorber film. See text for details

With the goal of obtaining clean surfaces by an UHV transfer, a growth system was setup and connected to the KPFM [20]. The system is similar to the previously used PVD system with three elemental sources for Cu, Ga and Se. However, during deposition the pressure remains below 10^{-8} mbar (base pressure 10^{-10} mbar). After a vacuum transfer of the CuGaSe₂ absorber to the UHV-KPFM we measured the surface morphology (Fig. 10.3a and b) electronic properties. The morphology corresponds well with the one obtained previously (see Fig. 10.1), showing grains with sizes of the order of 0.5–1 μm. The additionally measured work function image shows a fairly homogenous

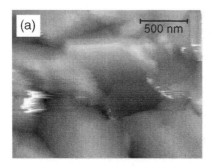
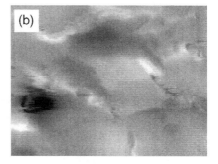

Fig. 10.3. Surface of a UHV-PVD-grown CuGaSe₂ film: (**a**)topography (*gray scale* corresponds to height differences of 320 nm); (**b**) simultaneously measured work function ($\Phi = 5.41 - 5.65$ eV). Reprinted from [20] with permission

value throughout the surface with $\Phi = 5.53 \pm 0.12\,\text{eV}$. This value is considerably higher than the one observed in Fig. 10.1. When the sample surface is illuminated (675 nm, 20 mW), the work function increases by ~80 meV. This reduction in surface band bending is likely due to charged surface states at the sample surface. As the sample surface is UHV-clean the surface states are not induced by contamination. Due to the illumination-induced flattening of the bands at the surface, the work function obtained under illumination is considered to reflect the distance from the vacuum level to the Fermi level in the bulk. Using the bandgap ($E_g = 1.68\,\text{eV}$) and the electron affinity ($\chi = 4.0\,\text{eV}$) of $CuGaSe_2$, we can estimate the position of the Fermi level to be $E_F - E_C = 75\,\text{meV}$, confirming the p-type of the material.

For a better comparison with the previous data, we also studied the effect of air exposure on the electronic surface properties. After exposure to air for 10 min, the sample was introduced in the UHV-KPFM and remeasured. The morphology remained unchanged (as expected) and also the contrast in the work function image did not change. However, the absolute value of the work function decreased considerably, showing a value of $\Phi_{\text{air}} = 4.66\,\text{eV}$ after air exposure. The value under illumination increases by only ~35 meV, indicating a Fermi-level position near midgap. In Fig. 10.4 we show a plot of the work function in the dark and under illumination depending on the surface treatment. The figure also includes results after an annealing treatment (110°C for 1 h) of the air-exposed sample. Also included are the results of the PVD-grown $CuGaSe_2$ sample from Fig. 10.1. Both the air-exposed samples show a similar work function. The lower work function after air exposure can be explained by a modified surface due to (1) a water film on the surface, (2) oxidization of the surface, or (3) other surface contamination.

Comparing the work function measured on the sample cleaned by sputterannealing and the UHV clean surface shows a difference of ~300 meV.

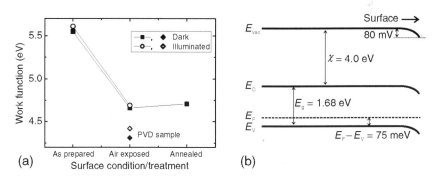

Fig. 10.4. (a) Dependence of the work function on the surface condition for the UHV-PVD samples (*squares and circles*) and PVD samples (*diamonds*); (b) schematic band diagram in the dark for the surface of UHV-PVD-grown $CuGaSe_2$. Reprinted from [20] with permission

The cleaning procedure of the surface can be one origin for this deviation, however, since the materials were grown in two different systems; also, inherent material properties (i.e., the doping concentration) can be responsible for this difference. Nevertheless, from the results presented it is concluded that samples transferred into the UHV-KPFM from air should be subjected to a cleaning process prior to work function determination.

10.3.2 Surface Orientation

As can already be seen in Figs. 10.1b and 10.3b, the work function images show well-defined areas of constant and distinct work function. This observation will be studied and discussed in more detail in this section [21]. For this purpose we studied an oriented $CuGaSe_2$ thin film grown on a ZnSe (1 1 0) surface by metal-organic chemical vapor deposition (MOCVD) [22]. The sample was transferred from the growth system under inert-gas atmosphere into the UHV-KPFM, thus avoiding unwanted exposure to oxygen and water. The topography and the work function of the MOCVD-grown $CuGaSe_2$ surface are shown in Fig. 10.5a and b, respectively. It is apparent that different crystal facets exhibit distinct values of the work function. From X-ray diffraction measurements the [220] direction was determined to be perpendicular to the sample surface. From the truly three-dimensional topographic information provided by NC-AFM, the crystal orientation of the different facets was indexed according to an analysis of the angles between the facets of single grains and the surface normal. Comparison of these angles with those expected for the (tetragonal) $CuGaSe_2$ crystal structure was used to assign the crystallographic orientation of the various facets. The result is included in Fig. 10.5b. It is well known that (1 1 2) surfaces develop preferentially during crystal growth [23]; they are therefore likely to be found frequently in the present sample.

The distinct work function values for different facets are due to a surface dipole that depends on the specific surface structure; the atomic arrangement at the surface, i.e., surface relaxation and reconstruction, depends on the crystal orientation of the facet. Therefore, the atoms/ions will form a surface dipole that is characteristic for the surface orientation [24]. The indexing of the (1 1 2) planes is as follows: the (1 1 2) plane is metal-terminated, whereas the (1 1 2̄) plane is Se-terminated. Using the Pauling electronegativities, the metal termination of the (1 1 2) surface results in a lower work function due to the surface dipole (as indexed in Fig. 10.5b). The abundance of the Se-terminated (1 1 2̄) facets is likely due to the Se-rich growth conditions.

10.3.3 Surface Inhomogeneities

As the growth process of the ternary compounds can be quite complicated, the possibility for the development of secondary phases during deposition is high. It is well known that deposition under Cu-rich growth conditions results in

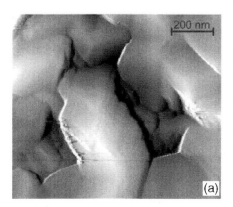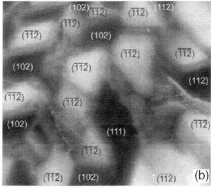

Fig. 10.5. KPFM measurement of the MOCVD-grown CuGaSe$_2$ thin film grown on single-crystalline ZnSe (1 1 0) substrate: (**a**) topography image shows distinct crystal facets on the (220) oriented CuGaSe$_2$ film (*gray scale = 384 nm*); (**b**) representation of the simultaneously measured work function ($\Phi = 4.85$–5.09 eV). The crystallographic orientation of the facets was assigned based on the angles to other facets and to the surface normal. Reprinted from [21] with permission

secondary phases like Cu$_x$S or Cu$_{2-x}$Se [25], which usually segregates at the top surface and can subsequently be removed by, for example, KCN etching. This etching is avoided when a Cu-poor growth or a Cu-poor final growth step is used.

With the aim of exploring whether secondary phases can be detected by KPFM, we studied the surface of a stoichiometric CuGaSe$_2$ film [26] deposited onto Mo-coated soda lime glass by CVD [27]. The sample was transferred in inert gas from the growth chamber into the UHV-KPFM to avoid surface contamination. The topography of the polycrystalline film with grain sizes on the order of 1 μm is shown in Fig. 10.6b. Due to height differences of up to 1 μm, scanning on the sample proved to be very difficult. The work function images in Fig. 10.6a in darkness and Fig. 10.6c under illumination (20 mW at 675 nm) show again the orientation-dependent work function for the various facets (see Sect. 10.3.2). Under illumination most regions of the sample show an increase of the work function by about 80 meV, as illustrated by the line scans in Fig. 10.6d. However, one grain in Fig. 10.6 (marked by ellipse) shows a strong negative SPV of about −200 mV (left end of the line scan). This part of the grain is possibly covered by Cu$_{2-x}$Se, an impurity phase that usually occurs during Cu-rich growth [28]. We have previously shown that an SPV of about −0.3 eV evolves for the segregation of a Cu$_{2-x}$Se layer on top of a CuGaSe$_2$ surface [28].

Since the absorbers used in chalcopyrite solar cells are of polycrystalline nature, it is apparent that grain boundaries might play a role in the solar cell device operation. The grain boundaries are structurally different from the bulk of the grains, which likely results in modified electronic properties.

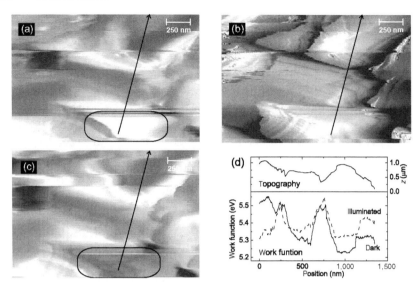

Fig. 10.6. KPFM measurements of CVD-grwn CuGaSe$_2$ on Mo/glass: **(a)** work function image ($\Phi = 5.14 - 5.53\,\text{eV}$) of the surface in the dark **(b)** simultaneously measured topography image (gray scale $\approx 1044\,\text{nm}$) **(c)** work function image of the surface under illumination ($\phi = 5.21 - 5.53\,\text{eV}$) **(d)** line scans of topography and work function. Reprinted form [26] with permission

In a recent study using electronic Hall measurements on PVD-grown CuGaSe$_2$, Schuler et al. [29] interpreted their results by a potential barrier for majority carrier transport located at the grain boundaries. They determined a band bending of $\Delta\Phi_{\text{gb}} = 60 - 135\,\text{meV}$ induced by charged trap states at the grain boundaries, with a density of $n_{\text{t}} = 1.1 - 1.4 \times 10^{12}\,\text{cm}^{-2}$.

As KPFM is capable of resolving potential variations on the surface with a nm resolution, we attempted to image the potential change at individual grain boundaries directly [19]. We studied samples from the same PVD process [18]; prior to introduction into the UHV-KPFM the sample was briefly exposed to air. The topography of the sample surface is shown in Fig. 10.1a. In the work function image in Fig. 10.1b, a dark "ring" around the individual grains is observed, indicating a work function decrease at the grain boundaries. To quantify this effect, we extract a representative linescan along the line in Fig. 10.1b. In Fig. 10.7, the work function (solid squares) is seen to drop at the grain boundaries by ~100 and $\sim170\,\text{meV}$. The analysis of a number of grain boundaries results in an average of $110 \pm 24\,\text{meV}$. The open circles show the work function along the same line for the sample measured under illumination. The overall increase $\sim110\,\text{meV}$ of the work function is due to a reduced band bending at the surface. Additionally, the work function decrease at the grain boundaries is reduced, giving an average value of $76 \pm 30\,\text{meV}$ under illumination.

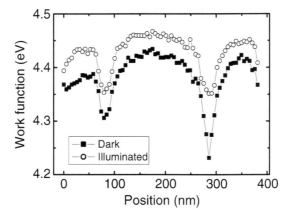

Fig. 10.7. Linescan along the white line in Fig 10.1b showing the decrease of the work function at the grain boundaries. *Solid squares* measured under dark conditions and *open circles* represent the work function along the same line under illumination (20 mv, $\lambda = 675\,\mathrm{nm}$). Reprinted from [19] with permission

Besides the size of the potential drop at grain boundaries, we can also extract information about the width of the space charge region (SCR) developing at the grain boundaries. We find an average SCR width of $42 \pm 9\,\mathrm{nm}$ under dark conditions and $56 \pm 10\,\mathrm{nm}$ under illumination, which is more or less similar, considering the experimental uncertainty.

Besides the selenide material CuGaSe$_2$, we also studied a sulfide material with a similar bandgap, i.e., Cu(In,Ga)S$_2$. The sample was prepared by PVD with a Ga content of $\sim$5%, leading to $E_g \sim 1.55\,\mathrm{eV}$ [30]. Height variations in topography (not shown) are around 500 nm and in the work function image (not shown) we again observe a decrease at the grain boundaries. Quantitative evaluation of a large number of grain boundaries results in an average band bending of $117 \pm 26\,\mathrm{meV}$; the SCR width is found to be $82 \pm 13\,\mathrm{nm}$.

Considering the charges located at the grain boundary, the net doping P_{net} of the absorber material can be determined from the magnitude of the band bending $\Delta\Phi_{\mathrm{gb}}$ and the SCR width w [17]:

$$P_{\mathrm{net}} = \frac{2\varepsilon_0\varepsilon\,\Delta\Phi_{\mathrm{gb}}}{e^2 w^2}, \tag{10.5}$$

where $\varepsilon = 11$ [31] is the dielectric constant of the absorber material, ε_0 the vacuum permittivity and e the elementary charge. For the PVD-grown CuGaSe$_2$ the net doping results in $P_{\mathrm{net}} \approx 9 \times 10^{16}\,\mathrm{cm}^{-3}$ and for Cu(In,Ga)S$_2$ in $P_{\mathrm{net}} \approx 3 \times 10^{16}\,\mathrm{cm}^{-3}$. SCR is small compared to the grain size, therefore it does not extend to the center of the grains. Thus, the density of charged trap states p_{gb} can be estimated according to a model originally proposed for transport across grain boundaries in polycrystalline silicon [32]:

$$p_{\mathrm{gb}} = \frac{1}{e}\sqrt{8\varepsilon_0 \varepsilon P_{\mathrm{net}} \Delta \Phi_{\mathrm{gb}}}. \tag{10.6}$$

In the present case the density of charged trap states results in $p_{\mathrm{gb}} \approx 8 \times 10^{11}\,\mathrm{cm}^{-2}$ and $p_{\mathrm{gb}} \approx 4 \times 10^{11}\,\mathrm{cm}^{-2}$ for CuGaSe$_2$ and Cu(In,Ga)S$_2$, respectively.

The results obtained for CuGaSe$_2$ compare well with the transport measurements by Schuler et al. [29], who found $P_{\mathrm{net}} \approx 1 - 5 \times 10^{17}\,\mathrm{cm}^{-3}$ for the net doping concentration, $\Delta \Phi_{\mathrm{gb}} = 60\text{--}135\,\mathrm{meV}$ for the band bending at the grain boundaries and $p_{\mathrm{gb}} \approx 1 \times 10^{12}\,\mathrm{cm}^{-2}$ for the charged defect density at the grain boundaries. It should be pointed out that two different techniques are compared in this case. Whereas KPFM allows analysis of a number of individual grain boundaries, Schuler et al. investigated an "average lowest" grain boundary barrier, determined by the average barrier of the percolation path through the material. Nevertheless, the results of this study and the present KPFM study are in good agreement.

From the present results, no significant difference in the electronic properties at the grain boundaries between the selenide material CuGaSe$_2$ and the sulphide material Cu(In,Ga)S$_2$ is observed. As both materials are p-type, it is expected that the bands are bent due to defects located at grain boundaries. The origin of the defect states at the grain boundaries can be manifold, for example, dangling bonds, oxygen contamination due to air exposure, Na contamination from the glass substrate, etc. [5].

10.3.4 Cross-Sectional Studies on Complete Devices

The properties of a complete chalcopyrite solar cell device can be interpreted in the form of its band diagram, which contains information about the band alignment from one layer to the next, as well as SCRs. Klenk [33] discussed the influence of various situations in the band diagram on the solar cell performance. As charge carrier recombination at the absorber–buffer interface is to be avoided, the conduction band should be "brought" close to the Fermi level at this position, for example, by an inverted interface of the p-type absorber, by using an n$^+$ window and/or by an appropriate amount of interface charge. In order to avoid a reduced barrier for recombination and V_{oc} losses, a cliff in the band alignment between buffer and absorber is also to be avoided. As the physical information contained in a band diagram is of importance to the understanding of solar cell devices, it is naturally desirable to obtain this information experimentally. In an approach to this task, we developed a method to study the cross-section of a complete solar cell device. As the KPFM measures the local work function, i.e., the distance between the local vacuum level and the Fermi level, the obtained results are useful for the construction of a band diagram.

For the initial study a CuGaSe$_2$ device was used [35]. The absorber layer was deposited onto Mo-coated soda lime glass by CVD from two sources, Cu$_2$Se and Ga$_2$Se$_3$, using I and HCl as transport gases, respectively [27].

A two-stage growth process deposits an initial Cu-rich layer, which is converted to an overall Ga-rich composition by a second step using only the Ga_2Se_3 source. On top of the absorber a 50-nm-thick CdS buffer was deposited by chemical bath deposition, followed by a sputtered two-layer ZnO window consisting of 80 nm undoped and 400 nm Ga-doped ZnO, completing the heterostructure. The cell parameters determined by current–voltage measurements under AM1.5 illumination are: $V_{oc} = 820$ mV, $J_{sc} = 10.6$ mA cm^{-2}, FF $= 54\%$, $\eta = 4.6\%$.

Cleavage of chalcopyrite solar cells proved to be very difficult due to the layered structure of various materials [34]. Instead, a device is cut into two halves and glued face-to-face using a conductive, UHV-suitable glue. In some cases a thin gold or copper stripe is introduced in the middle between the devices, serving as a reference for work function calibration. This sandwich is then cut to expose the cross-section of the devices and polished using alumina paste with a final grain size of $\sim$20 nm. Thus, a flat surface exposing various layers is obtained, which is well suited for AFM characterization.

Measuring the sample immediately after introduction into the UHV-KPFM results in a reduced CPD contrast of $\sim$100 meV between the different layers. A similar result is obtained after annealing the sample in the UHV system (for 1 h at 110°C). In both cases, illumination did not result in a change of the work function, indicating Fermi-level pinning at the contaminated surface. Therefore, the sample was subsequently cleaned by several soft sputtering processes using Ar-ions (500 eV) under 45° incident angle.

An overview of the work function values of the different materials after the various cleaning steps is presented in Fig. 10.8 (solid symbols). Open symbols represent the values under illumination. The work function of the $CuGaSe_2$ layer increases under illumination; illumination reduces the band bending at the p-type semiconductor surface achieving nearly flat band conditions at the

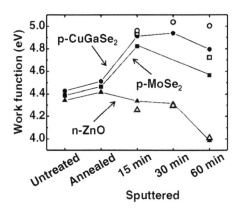

Fig. 10.8. Work function of the different materials in a CVD-grown CuGaSe₂ solar cell after different surface treatments. Open symbols represent the values under super bandgap illumination ($\sim$60 mW, $\lambda = 442$ nm). Reprinted from [35] with permission

surfaces. Under these conditions and after 15 min Ar-sputtering, the work function of CuGaSe$_2$ is 4.96 eV. For the highly n-doped ZnO we determine a work function of 4.26 eV, in reasonable agreement with results from Moormann et al. [36]. Up to a sputtering time of 30 min only the work function of the p-type semiconductors changes significantly. Due to the high carrier concentration of ZnO ($n \approx 10^{21}$ cm^{-3}) the effect of adsorbates on the position of the surface Fermi level should be minimal. For longer sputtering times a significant reduction of the work function of all layers is observed. As some elements are preferentially sputtered, this is attributed to a change in the surface stoichiometry. Otte et al. [37] have shown by XPS and Raman studies on sputtered CuGaSe$_2$ samples that metallic Ga ($\Phi = 4.2$ eV) is found on the CuGaSe$_2$ surface after Ar-ion etching for 35 min at 500 eV.

We therefore analyze the cross-sections after 15 min Ar-ion sputtering. The topography and work function under dark conditions of the Ga-rich CuGaSe$_2$ cell are shown in Fig. 10.9a and b, respectively [35]. Figure 10.9c shows the work function image under illumination. Due to different sputtering rates of the materials a topographical contrast between the layers develops. From the bottom to the top, the Mo back contact, CuGaSe$_2$ absorber, and ZnO window layer are seen. In the work function image the different layers show a contrast of up to 570 meV under dark conditions and up to 700 meV under illumination. Between the absorber and the Mo back contact an additional layer of about 100 nm thickness with a distinct work function is observed. This is attributed to an MoSe$_2$ intermediate layer, formed during the CVD absorber growth. High-resolution transmission electron microscopy and scanning energy dispersive X-ray detection confirm the presence of an interfacial MoSe$_2$ layer with a

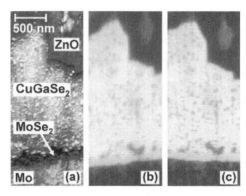

Fig. 10.9. Cross-sectional UHV-KPFM images ($1,200 \times 2,800$ nm^2) of a CVD-grown CuGaSe$_2$ solar cell after 15 min Ar-ion sputtering (500 eV): (**a**) topography (*gray scale* = 45 nm); (**b**) work function ($\Phi = 4.44 - 5.01$ eV); and (**c**) work function ($\Phi = 4.48 - 5.18$ eV) under illumination ($\sim$60 mW, $\lambda = 442$ nm). Reprinted from [35] with permission

thickness of 150 nm [38]. The Mo–MoSe$_2$ interface exhibits an intimate contact formed during the deposition process. For the MoSe$_2$–CuGaSe$_2$ interface we observe some voids in the interface, related to a nonhomogenous nucleation process of the crystallite seeds during the CuGaSe$_2$ deposition. The role and possible effects of the MoSe$_2$ layer in CVD-grown CuGaSe$_2$-based solar cells are under current investigation [39].

The above example used a solar cell with a Ga-rich absorber layer. This material results in the best solar cell performance. In order to shed light on the question, what details are responsible for the failure of solar cells with Cu-rich CuGaSe$_2$, we also analyzed a cell based on a Cu-rich CuGaSe$_2$ absorber [39]. The sample growth followed the same procedure described above. Only the second step in the growth process, converting the Cu-rich layer into a final overall Ga-rich, composition was omitted. Subsequently, the cell was completed without any etching treatment (as normally required) and the cross-section prepared as described above. A work function image of the Cu-rich CuGaSe$_2$ solar cell device is shown in Fig. 10.10a. In the left upper corner the glue is seen, followed by the ZnO window with the low work function,

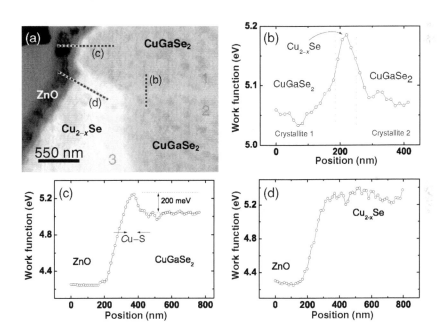

Fig. 10.10. KPFm measurement of a Cu-rich CVD-grown CuGaSe$_2$ solar cell device cross-section; (**a**) work function image showing the several materials and phases; (**b**) linescan showing the presence of a Cu$_{2-x}$Se foreign phase at the grain boundary between two CuGaSe$_2$ grains; (**c**) transition from ZnO window to the CuGaSe$_2$ with the electronic barrier due to the CuS phase; and (**d**) transition from the ZnO to the Cu$_{2-x}$Se gain showing no barrier [39]

the buffer layer and the absorber material on the right-hand side. In conjunction with scanning energy dispersive X-ray emission measurements on the same sample position, various phases as indicated were assigned. Grains "1" and "2" were identified to be CuGaSe$_2$, whereas grain "3" was found to be Cu$_{2-x}$Se. This finding is also supported by the KPFM measurement, where the work function of grain "3" is considerably higher than that of CuGaSe$_2$ (grains "1" and "2"). At the interfaces between CuGaSe$_2$ and ZnO and between the CuGaSe$_2$ grains interfacial phases with distinct work function are observed, as highlighted in Fig. 10.10b and c. The former linescan runs across two grains, including a grain boundary. Contrary to the band bending effect described in Section 10.3.3, the work function increases at the present grain boundary. This is explained by the presence of an interfacial foreign phase, namely a Cu$_{2-x}$Se phase. Apparently, this Cu-rich phase does not only segregate at the absorber surface during growth [25], but also forms along the grain boundaries. The work function clearly indicates its p$^+$ character, forming a conductive channel to the back contact. Thus, the cell is shorted by this secondary phase. The other interfacial phase observed between absorber and window layer (highlighted in Fig. 10.10c) also deserves some attention. This is the location of the buffer layer, which in the regular solar cell consists of a chemical bath deposited CdS layer. However, as confirmed by high-resolution TEM images, in the present sample the interface between absorber and window consists of a thin CuS layer. Apparently the Cu$_{2-x}$Se surface layer induces the growth of a CuS phase rather than the CdS buffer layer. A detrimental effect of this layer in addition to its nearly metallic character is seen from the KPFM measurement illustrated in Fig. 10.10c. The CuS phase has a work function of ∼200 meV higher than the CuGaSe$_2$ grain. This represents a barrier for minority carrier transport from the absorber into the window, which might be detrimental for solar cell performance. This effect could be avoided by a proper etching treatment of the absorber, removing the Cu$_{2-x}$Se phases from the surface prior to the buffer deposition. However, the initially described Cu$_{2-x}$Se at the grain boundaries will likely not be removed by wet chemical etching, since they are confined to the inside of the absorber. As shown in Fig. 10.10d, no barrier is observed at the transition of the Cu$_{2-x}$Se to the ZnO window.

In another study on a solar cell cross-section the impact of a damp heat (DH) treatment was addressed. The DH test is designed to obtain information about ageing of solar modules. For this purpose the unencapsulated device is exposed to DH condition at 85°C and 85% relative humidity for a period of 100 h. The degradation during this test should be kept minimal as during the life cycle of a solar panel its operating efficiency should be as constant as possible. The effect of this DH treatment on solar cells based on Cu(In,Ga)(S,Se)$_2$ absorbers (provided by Shell Solar) was investigated by several different techniques [40]. The cell contained a ZnO window extension layer (WEL) instead of the regular CdS buffer. This ZnO-WEL was grown using an ion layer gas reaction (ILGAR) process [41]. From electronic device characterization a degradation from 13.8% efficiency to 9.3% was found, mainly due to losses in the open-circuit voltage (decrease from 570 mV to 490 mV) and fill factor (decrease from 71% to 54%). However, quantum efficiency measurements gave indications that the collection length in the device is extended due to the DH treatment.

We tried to obtain direct evidence for the change in the collection length from KPFM measurements. A cross-section according to the above description was prepared for the cell exposed to DH conditions (post-DH) and a second cross-section using a cell without DH treatment (pre-DH) as a reference. Figure 10.11 shows the CP image of the sample before DH (a) and after DH (b) test. The RF-sputtered ZnO/Ga window layer shows a low CP corresponding to highly n-doped material. The absorber on the right side of the image has a high CP, according to p-type material. (The thin ILGAR-ZnO WEL between emitter and absorber cannot be resolved.) In the sample before DH (Fig. 10.11a), part of the glue is visible on the left-hand side of the image. Corresponding line scans are shown in Fig. 10.11c and d, in which also the potential drop from the absorber to the RF n^+-ZnO window and the width of the SCR are given. Evaluation of a number of line scans results in average values for the SCR before DH of $SCR_{preDH} = 275 \pm 80$ nm and after DH of $SCR_{post\text{-}DH} = 435 \pm 90$ nm. Thus, the qualitative observation of the larger collection length after DH of the WEL device is supported in the direct observation of an increased SCR width by KPFM, assuming no significant degradation in the minority carrier diffusion length. The observed

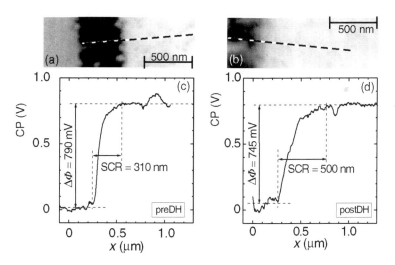

Fig. 10.11. Results of the KPM measurements on an ILGAR deposited ZnOWEL/Cu(In,Ga)(S,Se)$_2$ solar cell cross-section. The contact potential is shown for a device before (pre-) **(a)** and after (post-) **(b)** damp heat treatment. Line scans are given before **(c)** and after **(d)** DH including quantitative information about built-in voltage and space charge region width

potential drop across the pn-junction is related to the built-in voltage V_{bi} of the device. The analysis of a set of representative line scans results in mean values of $V_{\mathrm{bi}} = 800 \pm 100$ mV for the sample both before and after the DH test. Thus, the DH-induced deterioration of V_{oc} is within the uncertainty of the V_{bi} determination.

The results presented on the cross-sectional device characterization clearly show the power of the KPFM method in solar cell device characterization. The presence of minute segregations of foreign phases and their electronic relevance can be detected. However, care has to be taken with quantitative interpretation of the results. As the information is obtained on a surface, modification due to surface contamination, damage (from polishing and/or sputtering) and surface band bending has to be taken into account before any quantitative evaluation. Likely, these issues will have to be carefully discussed for each individual measurement, limiting the application of this analysis technique to selected case studies.

10.4 Conclusion and Outlook

With KBFM a powerful method for material characterization on the nanometre scale has been introduced, providing valuable electronic information about semiconductor materials and devices. Various applications of this method to chalcopyrites have been presented. Operation in UHV allows quantitative determination of the work function and processes were introduced to clean contaminated surfaces prior to KPFM investigation. On the surface of polycrystalline chalcopyrite thin films a variety of inhomogeneities were observed. The facets of grains show distinct work function, possibly leading to a lateral variation of the band alignment from the absorber to the subsequent buffer/window layer. Foreign phases can be detected by virtue of their distinct work function. Electronic effects have been observed for individual grain boundaries, showing a downward band bending due to charged trap states. This band bending was observed to decrease under illumination; thus, the situation is improved under solar cell working conditions. The application of the technique to a cross-sectional arrangement of complete devices provided information about foreign phases inside the absorber and on its surface. Valuable electronic information for the construction of band diagrams was obtained, including, for example, the SCR width. Difficulties and obstacles in the quantitative interpretation of results were discussed.

Further application of KPFM in chalcopyrite-based solar cell research will certainly prove to be valuable. It is well suited for the investigation of specific problems and can provide answers inaccessible by other techniques. Besides the ones presented, future applications might aim for additional properties. One possibility could, for example, be the measurement of a solar cell device under applied bias. Definitely more time will be devoted to sample surfaces prepared and transferred under UHV conditions. These surfaces should serve as a reference, to be compared to the samples obtained from technologically important growth processes. By this comparison, effects inherent to materials should be distinguishable from "artefacts" induced by surface contamination or oxidation.

Acknowledgments

The author would like to express his thanks to S. Schuler, S. Nishiwaki, R. Kaigawa, A. Meeder, D. Fuertes Marrón, M. Bär, M. Rusu and Shell Solar for providing samples. P. Schubert-Bischoff is thanked for the preparation of the cross-sectional samples. The author is especially grateful to Th. Glatzel, H. Steigert, K. Ishii and N. Barreau for support and discussion in the KPFM group.

References

1. Ramanathan, K., Contreras, M.A., Perkins, C.L., Asher, S., Hasoon, F.S., Keane, J., Young, D., Romero, M., Metzger, W., Noufi, R., Ward, J., Duda, A.: Properties of 19.2% efficiency ZnO/CdS/CuInGaSe$_2$ thin-film solar cells. Prog. Photovolt.: Res. Appl. 11, 225–230 (2003)
2. Siebentritt, S.: Thin Solid Films 403-404, 1 (2002)
3. Schock, H.W., Noufi, R.: Prog. Photovolt.: Res. Appl. 8, 151 (2000)
4. Herberholz, R., Nadenau, V., Rühle, U., Köble, C., Schock, H.W., Dimmler, B. Sol. Energy Mater. Sol. Cells 49, 227 (1997)
5. Scheer, R.: Trends Vac. Sci. Technol. 2, 77 (1997)
6. Wiesendanger, R.: Scanning Probe Microscopy and Spectroscopy: Methods and Applications. Cambridge Univiversity Press, Cambridge (1994)
7. Binnig, G., Quate, C.F., Gerber, Ch.: Phys. Rev. Lett. 56, 930 (1986)
8. Weaver, J.M.R., Abraham, D.W.: J. Vac. Sci. Technol. B 9, 1559 (1991)
9. Nonnenmacher, M., O'Boyle, M.P., Wickramasinghe, H.K.: Appl. Phys. Lett. 58, 2921 (1991)
10. Vatel, O., Tanimoto, M.: J. Appl. Phys. 77, 2358 (1995)
11. Chavez-Pirson, A., Vatel, O., Tanimoto, M., Ando, H., Iwamura, H., Kanbe, H.: Appl. Phys. Lett. 67, 3069 (1995)
12. Fujihira, M.: Ann. Rev. Mater. Sci. 29, 353 (1999)
13. Sadewasser, S., Glatzel, Th., Shikler, R., Rosenwaks, Y., Lux-Steiner, M.Ch.: Appl. Surf. Sci. 210, 32 (2003)
14. Albrecht, T.R., Grütter, P., Horne, D., Rugar, D.: J. Appl. Phys. 69, 668 (1991)
15. Sommerhalter, Ch., Matthes, Th.W., Glatzel, Th., Jäger-Waldau, A., Lux-Steiner, M.Ch.: Appl. Phys. Lett. 75, 286 (1999)
16. Sommerhalter, Ch., Glatzel, Th., Matthes, Th.W., Jäger-Waldau, A., Lux-Steiner, M.Ch.: Appl. Surf. Sci. 157, 263 (2000)
17. Kronik, L., Shapira, Y.: Surf. Sci. Rep. 37, 1 (1999)
18. Schuler, S., Nishiwaki, S., Dziedzina, M., Klenk, R., Siebentritt, S., Lux-Steiner, M.Ch.: Mater. Res. Soc. Symp. Proc. 668, H5.14.1 (2001)

19. Sadewasser, S., Glatzel, Th., Schuler, S., Nishiwaki, S., Kaigawa, R., Lux-Steiner, M.Ch.: Thin Solid Films 431–432, 257 (2003)
20. Sadewasser, S., Ishii, K., Glatzel, Th., Lux-Steiner, M.Ch.: In: Fuyuki, T., Strunk, H.P., Werner, J.H., Sameshima, T. (eds.) Polycrystalline Semiconductors VII—Bulk Materials, Thin Films, and Devices, p. 319. Scitech Publ., Uettikon am See, Switzerland (2003)
21. Sadewasser, S., Glatzel, Th., Rusu, M., Jäger-Waldau, A., Lux-Steiner, M.Ch.: Appl. Phys. Lett. 80, 2979 (2002)
22. Bauknecht, A., Siebentritt, S., Albert, J., Lux-Steiner, M.Ch.: J. Appl. Phys. 89, 4391 (2001)
23. Jaffe, J.E., Zunger, A.: Phys. Rev. B 64, 241–304 (2001)
24. Mönch, W.: Semiconductor Surfaces and Interfaces. Springer Series in Surface Science, vol. 26.;pringer, Berlin Heidelberg New York (1993)
25. Klenk, R., Walter, T., Schock, H.W., Cahen, D.: Adv. Mater. 5, 114 (1993)
26. Sadewasser, S., Glatzel, Th., Rusu, M., Meeder, A., Fuertes Marrón, D., Jäger-Waldau, A., Lux-Steiner, M.Ch.: Mater. Res. Soc. Symp. Proc. 668, H5.4.1 (2001)
27. Fischer, D., Dylla, T., Meyer, N., Beck, M.E., Jäger-Waldau, A., Lux-Steiner, M.Ch.: Thin Solid Films 387, 63 (2001)
28. Sommerhalter, Ch., Sadewasser, S., Glatzel, Th., Matthes, Th.W., Jäger-Waldau, A., Lux-Steiner, M.Ch.: Surf. Sci. 482–485, 1362 (2001)
29. Schuler, S., Nishiwaki, S., Beckmann, J., Rega, N., Brehme, S., Siebentritt, S., Lux-Steiner, M.Ch.: In: Proceedings of the IEEE Conference, New Orleans, 504 pp, 2002
30. Kaigawa, R., Neisser, A., Klenk, R., Lux-Steiner, M.Ch.: Thin Solid Films 414, 266 (2002)
31. Syrbu, N.N., Bogdanash, M., Tezlevan, V.E., Mushcutariu, I.: Physica B 229, 199 (1997)
32. Seto, J.Y.W.: J. Appl. Phys. 46, 5247 (1975)
33. Klenk, R.: Thin Solid Films 387, 135 (2001)
34. Ballif, C., Moutinho, H.R., Hasoon, F.S., Dhere, R.G.,;l-Jassim, M.M.: Ultramicroscopy 85, 61 (2000)
35. Glatzel, Th., Fuertes Marrón, D., Schedel-Niedrig, Th., Sadewasser, S., Lux-Steiner, M.Ch.: Appl. Phys. Lett. 81, 2017 (2002)
36. Moormann, H., Kohl, D., Heiland, G.: Surf. Sci. 100, 302 (1980)
37. Otte, K., Lippold, G., Hirsch, D., Schindler, A., Bigl, F.: Thin Solid Films 361–362, 498 (2000)
38. Würz, R., Fuertes Marrón, D., Meeder, A., Rumberg, A., Babu, S.M., Schedel-Niedrig, Th., Bloeck, U., Schubert-Bischoff, P., Lux-Steiner, M.Ch.: Thin Solid Films 431–432, 298 (2003)
39. Fuertes Marrón, D.: PhD Thesis, Freie Universität, Berlin (2003)
40. Bär, M., Rusu, M., Reiß, J., Glatzel, Th., Sadewasser, S., Bohne, W., Strub, E., Muffler, H.J., Lindner, S., Röhrich, J., Niesen, T.P., Karg, F., Lux-Steiner, M.Ch., Fischer, Ch.H.: In: Proceedings of the 3rd World Conference on Photovoltaic Energy Conversion, Osaka, 2002
41. Bär, M., Muffler, H.J., Fische, Ch.H., Lux-Steiner, M.Ch.: Sol. Energy Mater. Sol. Cells 67, 113 (2001)

11

Interfaces of Cu-Chalcopyrites

A. Klein and T. Schulmeyer

A wealth of physical and chemical phenomena occur at interfaces of Cu-chalcopyrite semiconductors, which are important for thin film solar cells. Two important topics, the band alignment and Fermi-level-related defect formation at the chalcopyrite/II–VI semiconductor interfaces, will be presented in this chapter. Band alignment and defect formation might both contribute to the experimentally observed saturation of photovoltage of CuInSe$_2$ thin film solar cells with increasing bandgap and are discussed in relation to this effect. Results obtained using photoelectron spectroscopy with single crystals as well as with polycrystalline Cu(In,Ga)(S,Se)$_2$ thin films are presented and compared to each other.

11.1 Introduction

Interfaces generally play an important role in thin film solar cell devices. Band discontinuities and chargeable interface states (defects) determine the electrostatic potential distribution and thereby the rectifying properties and cell efficiency. The cell structure and a schematic energy band diagram for thin film solar cells with p-type Cu-chalcopyrites (Cu(In,Ga)(S,Se)$_2$, abbreviated here as CIS) as light absorbing material are given in Fig. 11.1. A buffer layer is inserted between the CIS absorber and the transparent ZnO window layer. Best efficiencies in excess of 19% are obtained with chemical bath deposited CdS buffer layers and a combination of nominally undoped (i) and a degenerately doped (n$^+$) ZnO window layer [1]. However, other buffer layer materials can also be used for high conversion efficiencies [2].

For best conversion efficiencies, it is implied that the Fermi level E_F varies as much as possible in the bandgap of the light absorbing material, which requires that E_F is as close as possible to the CIS conduction band at the CIS/buffer interface. This interface is therefore of crucial importance for the solar cell behavior.

The Fermi level position is determined by the charge distribution, which is given by the distribution of charged defect states and by the barrier heights

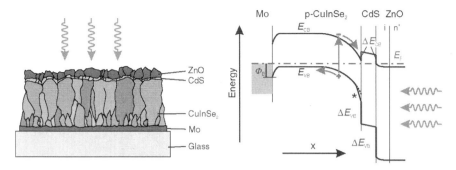

Fig. 11.1. Film sequence (*left*) and schematic band diagram (*right*) of a Cu(In,Ga)(S,Se)$_2$ (CIS) thin film solar cell. The barrier height at the Mo back contact (Φ_B) and at the CIS/CdS and CdS/ZnO interfaces (ΔE_{VB}, ΔE_{CB}), as well as the Fermi level position at the CIS/CdS interface are particularly important for device efficiency

at the interfaces. For the CIS/buffer heterointerface, these barrier heights are given by the discontinuities of the valence and conduction band edge, ΔE_{VB} and ΔE_{CB}, respectively. Understanding the properties of chalcopyrite solar cells thus requires knowledge about the band discontinuities and their dependence on processing conditions. Additionally, knowledge about charged interface states (defects) is required to fully identify the complete potential distribution.

11.2 Experimental Details

There are a number of techniques giving access to interfacial properties. Electrical analysis is highly sensitive to low defect concentrations, which is ideal for studying the important characteristics. However, these techniques in many cases require a completed device and it is often impossible to correlate the observed phenomena to a particular material or interface property. Photoelectron spectroscopy (PES), on the other hand, is a highly surface sensitive technique because of the low inelastic mean free path of the photo-excited electrons, which is of the order of a few nanometers [3]. With X-ray and UV excited PES (XPS and UPS, respectively) chemical and electronic properties of surfaces can be identified. The latter is due to the fact that binding energies of solid samples are measured with respect to the Fermi level of the system, which is the same in the investigated sample if electrical contact between the sample and the system is maintained. Moving the Fermi level in the bandgap of a semiconductor by doping or band bending therefore shifts all binding energies.

The interpretation of the Fermi-level position at a semiconductor surface is only meaningful for well-defined surface compositions, since already fractions of a monolayer of an adsorbate can induce considerable band bending.

This is the main reason why ultrahigh vacuum (UHV) environment is required for the study of semiconductor surfaces and interfaces. With a sticking coefficient of one, a surface is covered with one monolayer of an adsorbate within one second, when the partial pressure of the adsorbate is 10^{-6} Torr. Since a typical interface experiment requires several hours, a base pressure of the vacuum system of well below 10^{-9} Torr is necessary to maintain contamination-free surfaces. Good vacuum conditions are required not only during measurement but also during the complete experiment, which involves substrate surface preparation, film deposition and measurement. Substrate preparation and film deposition often require separate vacuum chambers not only because of limited space in the measurement chamber, but also to avoid contamination of the same. It is thus required to combine measurement chambers for surface analysis and those for substrate preparation and film deposition by an UHV transfer system.

A versatile solution is given by assembling different chambers in so-called integrated systems, where different chambers are "clustered" around a central sample distribution chamber. Such a system has been set up by the surface science group at the Darmstadt University of Technology. It combines a multitechnique surface analysis system (Physical Electronics PHI 5700) and several small multipurpose vacuum chambers. A schematic layout of the system is shown in Fig. 11.2.

A series of spectra obtained during an experiment performed in order to determine the valence band offset at a semiconductor heterojunction is shown

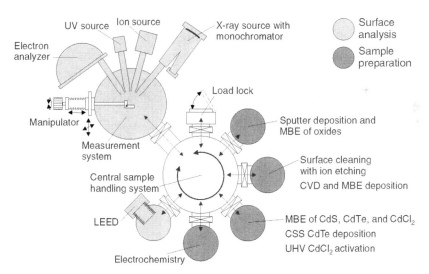

Fig. 11.2. Layout of an integrated surface analysis and preparation system for the study of thin film semiconductor interfaces. A photoelectron spectrometer is connected by a sample handling system to various deposition and surface treatment chambers. Preparation and analysis can be repeatedly performed under controlled ultrahigh vacuum conditions

in Fig. 11.3. The experiment starts with a clean substrate surface by measuring the binding energies of the corresponding core levels and of the valence band maximum.

Chemical and electronic changes are monitored during interface formation by stepwise deposition of the contact material onto the substrate. Deposition is typically finished after complete attenuation of the substrate emissions, which corresponds to about 5 nm film thickness if the film grows in a layer-by-layer mode. Core levels and valence band maximum are then determined for the film material. Since the energy distance between core levels and valence band maximum is a material constant, the valence band offset can be calculated from the energy difference between the core levels of substrate and film material, which is observed in the course of interface formation.

11.3 Band Alignment at Cu(In,Ga)(Se,S)$_2$/Buffer Interfaces

11.3.1 Single Crystal Cu(In,Ga)(S,Se)$_2$

Early experiments on the interface properties of Cu-chalcopyrites have been performed using single crystal samples, which were cleaved in UHV; [4, 5]. Such experiments might serve as reference experiments for the fundamental understanding of interface properties. A considerable part of the knowledge

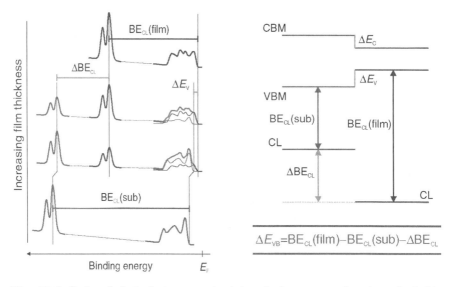

Fig. 11.3. Series of photoelectron spectra taken during an experiment conducted to determine the valence band offset ΔE_{VB} at a semiconductor heterojunction, measuring core levels (CL) and the valence band maximum (VBM)

on the physics and chemistry of semiconductor interfaces has been gained using photoemission experiments on single crystal Si or III–V semiconductor interfaces [6–8].

These studies have contributed to the understanding of Fermi-level pinning by systematically exploring a large number of material combinations in conjunction with theoretical models, which have introduced induced gap states. Compared to the well-known atomic structures of Si and III–V semiconductor surfaces [6], the understanding of the structure and electronic properties of single crystal CIS surfaces is still poor. Nevertheless, such experiments are necessary to compare the determined electronic interface properties with available theoretical calculations, which are provided by Zunger's group [9, 14].

Clean CIS single crystal surfaces are prepared by cleaving in UHV. For this, crystals have to be attached to a sample holder with the cleavage plane oriented perpendicular to the surface normal. For $CuInSe_2$ crystals, the cleavage plane was identified as the (0 1 1) plane [5]. Although oriented cleavage has also been achieved for $CuInS_2$ [4], the surface orientation could not be identified. In contrast, in our group it was not possible to cleave $CuGaSe_2$ along any crystal plane. Although only rugged surfaces were obtained by cleaving (fracture), these were free of contaminations and have therefore been used for interface studies [10].

An alternative way to prepare clean single crystal CIS surfaces for interface studies is provided by epitaxial growth, which can be achieved by either molecular beam epitaxy (MBE) [15–18] or metal–organic chemical vapor deposition (MOCVD) [19]. Such samples have also been used for interface studies, where sample transfer to the XPS system has been achieved either by mobile transfer chambers [11], or more recently by the decapping procedure [15], which will be described in detail in Sect. 11.3.2.

The band alignments determined by the measurements mentioned above are summarized in Fig. 11.4. To compare the valence bands offsets (VBOs) of II–VI semiconductors with different chalcopyrites with each other, the band alignment between the chalcopyrites needs to be known. As there exist no direct experimental determinations, one can refer to the theoretical values given by Zunger's group [9, 14]. The chalcopyrite valence bands in Fig. 11.4 are arranged according to these values. The obtained approximate transitivity of VBOs with CdS and ZnSe indicate that the experimental values agree with the calculated band offsets between the different chalcopyrites and suggest that the band alignment is determined by a bulk reference level as generally assumed [6–8]. The alignment of the chalcopyrite energy bands according to Fig. 11.4 is supported by an indirect determination of band alignment by assuming constant energy levels for bulk defects. This has very recently been presented by Turcu et al. [20] (see also Chap. 6). Such an alignment of energy bands according to defect levels is not peculiar to chalcopyrite semiconductors, but has also been discussed for III–V compounds [21–23].

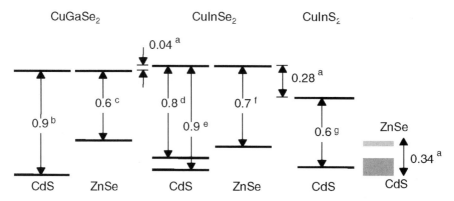

Fig. 11.4. Valence band maximum positions for interfaces between thermally evaporated CdS and ZnSe and single crystalline Cu-chalcopyrite semiconductors according to a: [9], b: [10], c: [11], d: [5], e: [12], f: [13], and g: [4]. Data from [9] (a) are valence band offsets calculated using density functional theory. All values are given in electronvolts

11.3.2 Thin Film Cu(In,Ga)Se₂

Surface Preparation

High conversion efficiencies of CIS solar cells have not been obtained with CIS single crystals, but only with polycrystalline thin film material. Furthermore, the deposition of the absorber film is important for device efficiency. It is thus imperative to analyze surfaces and interfaces of such films, which are also used for solar cell fabrication. Such films are typically obtained from deposition processes after long-term optimization procedures in dedicated deposition systems, which are often not suitable for integration in UHV systems. During optimization, the integration of the deposition chamber with surface and interface analysis is also typically not required.

To overcome these restrictions, a procedure for the "integration" of CIS thin film deposition into interface analysis systems has been successfully implemented. Thin CIS films were deposited in a deposition chamber at ZSW, Stuttgart, by the same process as used for high-quality solar cells (see also Chap. 12). Without breaking the vacuum, films were covered with a 1 μ thick amorphous Se layer after cooling of the sample below 100°C and before removing the sample from the vacuum system. This "capped" sample was sent to Darmstadt University by surface mail and introduced into the integrated system shown in Fig. 11.2. When the sample was heated at 300°C for 15 min, the Se layers evaporated because of the high partial pressure of Se, while the CIS layer, which was deposited at 550°C substrate temperature, remained intact. X-ray induced photoelectron spectra before and after heating of the sample are shown in Fig. 11.5. Initially, the surface only shows emissions from

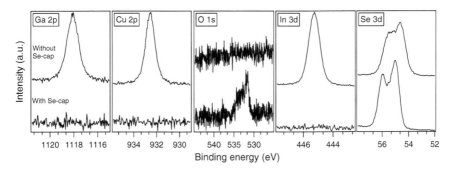

Fig. 11.5. X-ray induced photoelectron spectra of Se capped $CuIn_{0.7}Ga_{0.3}Se_2$ sample before (*bottom*) and after heating-off (*top*) of the Se cap layer

the Se cap layer, which is slightly oxidized. After "decapping", the sample shows all emissions from CIS and no more oxygen.

The stoichiometry determined by XPS using sensitivity factors supplied by the manufacturer of the spectrometer (Physical Electronics Phi 5700) corresponds to a Cu-poor $Cu(In_{1-x}Ga_x)_3Se_5$ (135) surface composition. We have found this surface composition for all investigated values of x (0.3, 0.7, and 1). As the films were grown under slightly Cu-poor conditions, this is in agreement with other measurements [24]. All thin film CIS materials used for studies described in this chapter are CIS films with slightly Cu-poor bulk composition, which expose Cu-poor surfaces after the decapping process.

Solar cells prepared with films from the deposition chamber used at the ZSW have efficiencies around 14%. To check whether the Se capping and decapping modifies the surface properties and thereby also the cell efficiencies, solar cells have been prepared at the ZSW by the standard cell fabrication process using decapped CIS layers. The current–voltage characteristics and quantum efficiency of the cell prepared from the decapped film are very close to those of a reference cell prepared from an identical absorber without the intermediate cap layer as shown in Fig. 11.6. Also the efficiencies of 13.9% (decapped CIS film) and of 13.2% (standard CIS film) are comparable. This result indicates that the capping/decapping sequence does not deteriorate the CIS surface, which opens wide opportunities not only for the use of capped CIS films in surface and interface studies, but also for process development.

Band Alignment

The VBOs have been determined between decapped $CuIn_{1-x}Ga_xSe_2$ ($x = 0.3$) and II–VI compounds CdX and ZnX (X = S, Se, Te). All II–VI semiconductors were deposited by thermal evaporation onto substrates kept at room temperature. In addition, $CuIn_{1-x}Ga_xSe_2$/CdS interfaces were investigated for $x = 0.7$ and 1. The binding energies of core-levels with respect to the valence band maxima, which have been used for the determination of the

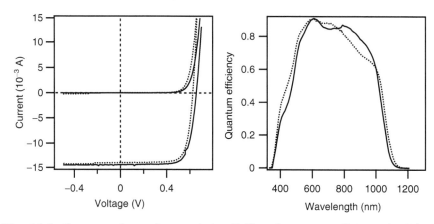

Fig. 11.6. Current–voltage characteristics (*left*) and quantum efficiency (*right*) of CIS solar cells prepared from an absorber, which was covered intermediately with an Se cap layer (*solid lines*) and an absorber without the Se cap layer (*dotted lines*)

VBOs are given in Table 11.1. Valence band maximum binding energies are determined using linear extrapolation of the leading edge of the valence band emissions and core-level binding energies are determined using least-squares fitting with Voigt profiles.

The experimentally determined VBOs are shown in Fig. 11.7 together with theoretical calculations from Zunger's group. Experimental values are given with ±0.1 eV error bars. Theoretical values are given for three different surface compositions: Cu(In,Ga)Se$_2$ (1 1 2), Cu(In,Ga)$_3$Se$_5$ (1 3 5), and Cu(In,Ga)$_5$Se$_8$ (1 5 8). These values are obtained using the transitivity rule and the calculated VBOs between the different CIS phases [9, 14, 27]. The applicability of the transitivity rule is supported by an experimental determination of the VBO at the CdS/CdTe interface. The value of $\Delta E_{VB} = 0.94$ eV [28] is very close to the one indicated by the data in Fig. 11.7. (0.97 eV).

11.3.3 Discussion

Several aspects of the band alignment at the CIS/buffer interface are important for the solar cell. In this section, the influence of bandgap and composition of the absorber will be mainly addressed. The first issue is of particular importance for CIS solar cells with large bandgaps.

It is well known that increasing the bandgap of CuInSe$_2$ by replacing In with Ga (or Se with S) does not lead to a corresponding increase of the open circuit voltage as theoretically expected [29, 30] (see also other chapters). One reason for this might be the change of the conduction band offset at the CIS/buffer interface from a spike to a cliff as indicated in Fig. 11.8 for the interfaces of single crystalline CuInSe$_2$ and CuGaSe$_2$ with CdS [5, 10, 12, 15].

Table 11.1. Core-level binding energies with respect to the valence band maxima for decapped $Cu(In_{0.7}Ga_{0.3})Se_2$ (CIS), CdX, and ZnX (X = S, Se, Te)

	CIS	CdS	CdSe	CdTe	ZnS	ZnSe	ZnTe
Cu $2p_{3/2}$	951.53	–	–	–	–	–	–
In $3d_{5/2}$	444.19	–	–	–	–	–	–
Ga $2p_{3/2}$	1117.10	–	–	–	–	–	–
Cd $3d_{5/2}$	–	403.48	403.86	404.39	–	–	–
Zn $2p_{3/2}$	–	–	–	–	1020.17	1020.75	1021.11
S $2p_{3/2}$	–	159.90	–	–	160.20	–	–
Se $3d_{5/2}$	53.52	–	52.59	–	–	52.86	–
Te $3d_{5/2}$	–	–	–	571.76	–	–	571.88

All values are in electronvolts

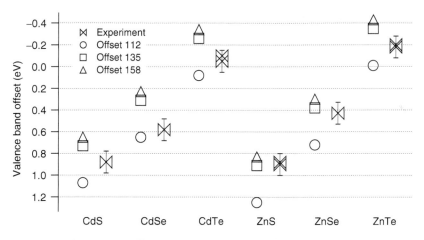

Fig. 11.7. Experimentally determined valence band offsets at decapped $CuIn_{0.7}Ga_{0.3}Se_2$/II–VI semiconductor interfaces [25, 26], compared to theoretical calculations for $CuInSe_2$ (1 1 2), $CuIn_3Se_5$ (1 3 5), and $CuIn_5Se_8$ (1 5 8) [9, 14, 27]

The VBOs observed with polycrystalline thin film material are almost identical to those obtained with single crystalline material, as shown in this chapter. However, the energy band diagrams will be different because of the different bandgaps due to the Cu-poor surfaces of the thin films; [31].

The VBOs at the $CuInSe_2$/CdS and $CuGaSe_2$/CdS interfaces are very similar. This is expected, since the valence bands of both materials are derived from Se 4p and Cu 3d states [32]. The similarity of the VBOs is also in agreement with theoretical calculations [14] and the experimentally determined defect energy level positions [20]. The difference in energy gap between $CuInSe_2$ and $CuGaSe_2$ therefore mainly affects the conduction band states. As PES can only detect occupied electronic states, the conduction band

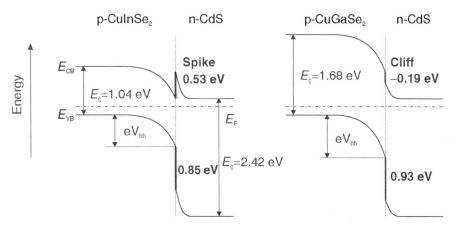

Fig. 11.8. Band diagrams for CuInSe$_2$/CdS and CuGaSe$_2$/CdS with experimental values for the valence band offset ΔE_{VB} of single crystalline CIS materials [5, 10, 12, 15]. The conduction band shows a spike ($\Delta E_{CB} > 0$) for CuInSe$_2$/CdS and a cliff ($\Delta E_{CB} < 0$) for CuGaSe$_2$/CdS. The presence of a cliff limits the band bending eV_{bb} in the substrate

offsets, which are more important for the CIS/buffer interface, cannot be determined using this technique. The bandgaps of the stoichiometric compounds have thus been used in Fig. 11.8 to give an idea about the conduction band offsets. This is, however, not strictly appropriate for high-efficiency CIS solar cells, which show a Cu-poor surface composition. The Cu-poor phases (135, 158, or others) exhibit larger bandgaps due to the smaller contribution of the Cu 3d states to the valence bands. These Cu 3d electrons lead to an upward shift of the valence band maximum with respect to the d-electron free case (e.g., ZnSe), called p–d repulsion [33]. In addition, intermixing at the interface can change the bandgaps of the absorber and also of the buffer layer as suggested by Morkel et al. [31].

A better description of the band alignment at the CIS/CdS interface in high-efficiency solar cells would be given by the VBOs of polycrystalline films as presented in Sect. 11.3.2. Based on the theoretical calculation of Zunger and coworkers, the VBO between the CIS 135 phase and II–VI semiconductors is expected to be 0.34 eV smaller than those between the corresponding 112 phase and the same II–VI compound [27]. However, while the VBO at the CuInSe$_2$/CdS interface has been determined as $\Delta E_{VB} = 0.8$–0.9 eV [5,12,15], we have repeatedly determined a value of $\Delta E_{VB} = 0.88 \pm 0.1$ eV for the Cu(In,Ga)$_3$Se$_5$ surface composition [25, 26]. The band alignments of single crystal and thin film CIS with CdS are compared in Fig. 11.9.

The energy diagrams shown in Fig. 11.9 indicate that the transitivity rule is not fulfilled for CIS/II–VI semiconductor interfaces if the Cu-poor surface phases are taken into account and if the theoretical band alignment between stoichiometric (1 1 2) and Cu-poor phases are inserted. Because of the simi-

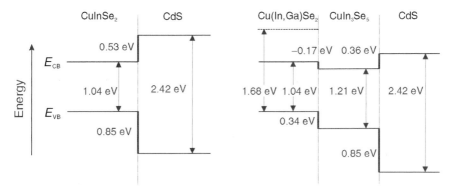

Fig. 11.9. Band alignment at CuInSe$_2$/CdS and CuIn$_3$Se$_5$/CdS according to the experimental results described in Sects. 11.3.1 and 11.3.2. Conduction band offsets are derived using bandgaps of 1.04 eV (1 1 2) and 1.21 eV (1 3 5) [27]. Band bending, Fermi-level positions and the influence of Ga on the valence band offset are omitted. A violation of the transitivity rule for semiconductor interfaces is evident

larity between the experimental VBO at the 112/CdS and 135/CdS interface, deviation from the transitivity rule just equals the offsets between 112 and 135 CIS, which is calculated as 0.34 eV [27].

Different explanations might account for this deviation from the transitivity rule: (1) different interfacial chemical bonds might lead to a modification of the charge transfer at the interface and thereby alter the interface dipole; (2) a chemical modification might occur during interface formation, which might result in a chemical composition of the interface that is independent of the substrate stoichiometry; (3) the different Cu concentrations detected using XPS might affect only the top layer of the material as recently suggested for anion-terminated (1 1 2) oriented surfaces of epitaxial CuInSe$_2$ films by Liao and Rockett [34].

At present we cannot conclude on either of these explanations, although we are in favor of the chemical modification model (2), since different interface dipoles would occur in the calculations as well and since a Cu-poor surface composition is not only evident from XPS stoichiometry determinations but also from less surface sensitive valence band spectra, which show a different density of states indicating less Cu contribution to the valence bands [35] (Schulmeyer, T. and Klein, A., unpublished results). Chemical modifications at the interface were also suggested to account for a slight deviation of the variation of the band offset at CIS/CdS with Ga content [26]. While theoretical calculations [14] and also the experimental determination of the band alignment via the defect energies [20] place the CuGaSe$_2$ valence band maximum 40 meV below that of CuInSe$_2$, the experimentally determined band offsets at Cu(In,Ga)Se$_2$/CdS interfaces show an opposite tendency and increase with Ga content independent of whether single crystal or decapped polycrystalline thin film material has been used [26]. Applying the transitivity rule thus suggests that the $CuGaSe_2$ valence band maximum is approximately 0.1 eV above that of $CuInSe_2$. The difference of the interface experiments compared to theoretical calculations and the alignment via defect energies is that the band offset determined by evaporation of CdS onto CIS corresponds to a truly interfacial property, which can be affected by a chemical modification of the interface. Although there is no clear evidence, it cannot be excluded that the difference is due to a different chemistry at the CIS/CdS interface depending on the Ga content. More systematic experiments need to be carried out to resolve the fundamental and important issue of chemical modifications at CIS/II–VI compound interfaces. A special case for interface modification will be presented in the Chap. 12.

11.4 Fermi-Level-Dependent Defect Formation

It has been known for some time that the noble metals Cu and Ag show high diffusivity in Cu-chalcopyrites (see [36] and references therein). Nevertheless, their diffusivity is usually neglected in the interpretation of solar cell device behavior. In the last years, however, the defect properties of CIGS have raised considerable attention [20, 27, 37–39] (see also other chapters in this book), partly stimulated by interface studies [10, 40, 41]. In Fig. 11.10, a number of valence band difference spectra for different CIS interfaces are shown. The differences were taken between partly covered surfaces and clean surfaces by shifting the spectra according to the overlayer induced band bending. Attenuation of the substrate intensity is accounted for by a multiplicative factor. In the case of undisrupted interfaces, the valence band spectra are superpositions of the clean substrate and overlayer spectra as indicated in Fig. 11.3. Subtracting suitably shifted and multiplied substrate spectra from spectra of the partly covered surface should therefore result in spectra corresponding to those of a thick overlayer. As indicated by the difference spectra in Fig. 11.10; this is the case only for the $CuInS_2$/CdS interface [4, 40]. The other interfaces show a considerable reduction of intensity at a binding energy of approximately 3 eV [10, 40, 41]. The corresponding energy states of the CIGS surfaces are the Cu 3d states, which are known from an experimental determination of the partial density of states using energy-dependent valence band spectra [32]. Consequently, the reduced intensity has been interpreted as a loss of Cu from the surface.

This observation can be related to the known mechanism of Fermi-level-dependent defect formation. The energy required to create a crystallographic point defect, which is associated with a positive or negative charge depends on the concentration of electrons, and hence on the Fermi-level position [27, 42]. Moving the Fermi level upwards increases the electron concentration in the conduction band and as a consequence may induce the formation of compensating defects, which are electron acceptors. Since these acceptors are able

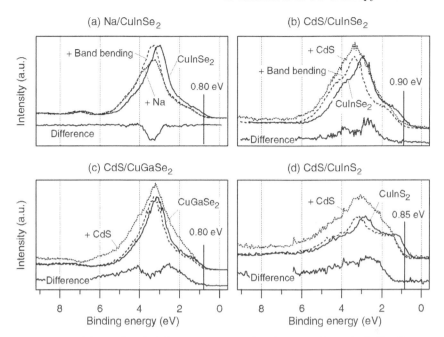

Fig. 11.10. Valence band difference spectra for CIS interfaces [40]. Binding energies are given with respect to the Fermi energy. Spectra from clean surfaces are represented by *solid lines*. The same spectra shifted by the induced band bending are indicated by *dashed lines*. Numbers correspond to the valence band maximum position in the substrate after band bending has occurred. Spectra with overlayers are represented by *dotted lines*

to bind the additional electrons, it is not possible to move the Fermi level beyond a certain limit, as this would always imply the formation of additional acceptors.

Since the Fermi-level position determines the electron (and hole) concentration, this limit determines a maximum concentration of electrons (or holes), generally referred to as *doping limit* [9,43]. The effect of acceptor (donor) formation by the upward (downward) movement of the Fermi level is known as *self-compensation*, which is also the topic of Chap. 7

In the spectra shown in Fig. 11.10, the Fermi level of the clean surfaces is close to the valence band maximum. In all cases, it moves upward in the bandgap with increasing deposition as a result of charge transfer between the substrate and overlayer or charged interface states. The surface band bending saturates at a Fermi-level position of $E_F - E_{VB} = 0.85 \pm 0.1\,\text{eV}$. This energy position agrees with the energy associated with the ionization energy of interstitial Cu in CuInSe$_2$, as calculated by Zhang et al. [27].

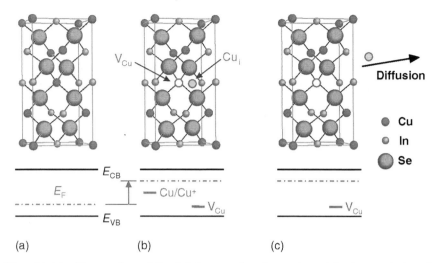

Fig. 11.11. Observation of Cu-deficient surface formation with photoemission. In the first step, the Fermi level is moved upwards in the bandgap by the adsorbate. The Fermi-level movement leads to the formation of interstitial Cu atoms, which can diffuse into the bulk of the CIGS absorber

The reduction of intensity from the Cu 3d valence states observed in the photoemission of CIGS interfaces is derived from a two-step process as sketched in Fig. 11.11. The clean surface corresponds to a Fermi-level position close to the valence band maximum, Fig. 11.11a. The increase of the Fermi level induced by adsorption, Fig. 11.11b, leads to an increase of electron concentration. The additional electrons are transferred to conduction band states. Since these are mainly derived from the Cu 4s orbitals, this leads to a reduction of Cu^+ to Cu^0 and hence to a destabilization of the Cu lattice bonds and to the formation of interstitial Cu_i. In the second step, Fig. 11.11c, Cu_i is removed from the surface by diffusion, induced by the concentration gradient of this species. It is the latter step, which enables the observation of the effect with the surface sensitive photoemission technique.

The transition from Cu^+ to Cu^0 by the increase of the Fermi level can also be illustrated using an electrochemical energy diagram [10]. In Fig. 11.12, an energy diagram for an interface between a $CuInSe_2$ surface and an electrolyte containing a Cu/Cu^+ redox couple is shown. The standard potential E_0 for the Cu/Cu^+ redox couple with respect to the standard hydrogen electrode is given as +0.52 V [44]. This corresponds to $E_0 \approx 5.0\,eV$ with respect to the vacuum level.[1] The electron affinity of $CuInSe_2$ amounts to $\chi = 4.6\,eV$ [41].

---

[1] To compare the electrode potentials with energy levels of solids, the potential of the standard hydrogen electrode has been determined with respect to the vacuum level yielding a value of 4.4–4.8 eV [45]. For convenience, the frequently used value of 4.5 eV has been used here.

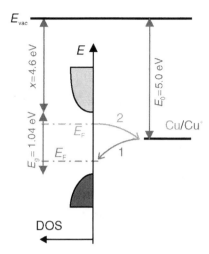

Fig. 11.12. Energy band diagram for an interface between CuInSe$_2$ and an electrolyte containing the Cu/Cu$^+$ redox couple. Cu is ionized if the Fermi level is below E_0. Increasing the Fermi level above E_0 leads to neutral Cu

If the Fermi energy is now below E_0, Cu is ionized. Increasing E_F above E_0 leads to neutralization of Cu.

It has been suggested that this model can be applied to explain the observed defect formation, when the chalcopyrites themselves are considered as solid electrolytes in equilibrium with the Cu/Cu$^+$ redox couple [10]. Although the use of electrochemical standard potentials to predict Cu defect formation is not established yet, the model presented here qualitatively explains several observations, since it relates the redox couples and the chalcopyrites on an absolute energy scale.

However, for more quantitative estimations of defect formation energies induced by electron transfer, the lattice stabilization energy (corresponding to the solvation energy in electrolytes) of interstitial Cu and lattice bound Cu$^+$ must be known in comparison to Cu in the metal and the solvation energy of Cu$^+$ in the electrolyte. The reorganization of the surrounding medium associated with defect formation is included in the theoretical calculations of defects in chalcopyrites, which have been performed by Zunger's group. They report a transition energy of $\approx 0.85\,\mathrm{eV}$ for Cu/Cu$^+$ in CuInSe$_2$ [27], which agrees excellently with the maximum Fermi-level position extracted from our experiments. Their calculations show that for Fermi levels above the Cu/Cu$^+$ transition energy, the energy required for creating an interstitial Cu atom does no longer increase with E_F, while the energy gain from creating Cu vacancies still increases. There is hence a net energy gain for the formation of a (V$_{Cu}$, Cu$_i$) defect pair for Fermi-level positions above the transition energy. The Cu/Cu$^+$ transition energy and the observed maximum Fermi-level position at different CIGS interfaces, further agree with the Fermi-level position at $Cu(In_{1-x},Ga_x)Se_2$ thin films, which has been determined using in situ photoemission measurements [30]. Throughout the whole composition range a consistent Fermi-level position of $E_F = 0.87 \pm 0.1\,eV$ is found at the absorber surfaces.

Arranging the different chalcopyrites on an absolute energy scale according to the calculated band alignment [9] explains the absence of Cu diffusion at the $CuInS_2/CdS$ interface (Fig. 11.10). The valence band maximum of $CuInS_2$ is approximately 0.2–0.3 eV below those of the selenides (see Fig. 11.4). If the involved defect levels have constant energies on the same absolute energy scale used for band alignment, as indicated by the measurements of Turcu et al. [20], the transition of charged to neutral Cu would occur for larger values of $E_F - E_{VB}$ in $CuInS_2$ as compared to $Cu(In,Ga)Se_2$. It is further suggested that Ag-containing chalcopyrites cannot be used for solar cell production, since the standard potential for Ag/Ag^+ of +0.8 eV is larger than the corresponding value for Cu/Cu^+. The formation of neutral Ag defect levels and the related doping compensation would consequently occur at a Fermi-level position $\approx$ 0.3 eV closer to the valence band edge.

An important consequence of the defect formation is related to the acceptor character of the V_{Cu} point defect [27]. As a consequence of acceptor formation, the band bending at the CIGS/CdS interface, which is important for the achievable photovoltage is limited by self-compensation. This is not a serious problem for low bandgapmaterials, since the achievable band bending is close to the bandgap. However, for increasing Ga content the valence band edge remains almost at a constant energy, while the conduction band shifts upwards in energy with increasing bandgap. Despite the larger bandgap, the band bending is still limited to the same maximum. The self-compensation mechanism will therefore lead to a saturation of photovoltage with increasing bandgap. This effect is independent of band alignment, which has also been suggested as a possible origin for photovoltage limitation (see Sect. 11.3).

11.5 Importance of Interfaces for V_{oc} Saturation

In this chapter, two aspects of the CIS/buffer layer interfaces have been presented, which might both be related to the experimentally observed saturation of the photovoltage of CIS thin film solar cells with increasing bandgap [29,30]: these are the issues of band alignment and of Fermi-level-induced defect formation (self-compensation). Concerning band alignment, it seems now quite well established that VBOs at the interfaces between chalcopyrites and II–VI semiconductors generally follow the transitivity rule. The experimentally observed differences in band alignment for the various II–VI compounds (see Sect. 11.3) are quite well reproduced by the values calculated by Zunger and coworkers [9,14]. In addition, the available VBO differences determined for $CuInSe_2$ and $CuInS_2$ are in accordance with these calculations, suggesting that the valence band maximum of CuInS$_2$ is 0.2–0.3 eV lower than that of CuInSe$_2$. In contrast to this generally converging picture on band alignments, the experimentally observed trends of ΔE_{VB} when replacing In with Ga are of opposite direction compared to the calculated behavior [26]. This has been attributed to differences in interface chemistry. However, although this has been observed for single crystalline and repeatedly also for polycrystalline material, the differences between experimental and calculated trend are below 200 meV. More and systematic experiments would therefore be required to better understand the differences in interface chemistry of CuInSe$_2$ and CuGaSe$_2$ with II–VI compounds.

So far, one of the most important aspects of band alignment at CIS/II–VI interfaces is still not resolved; i.e., the influence of Cu-poor surfaces. While theory suggests a difference of ΔE_{VB} of 0.34 eV between the 112 and the 135 composition, the observed VBOs are quite insensitive to absorber and interface preparation. Even for chemical bath deposited CdS layers, a VBO of $\Delta E_{VB} = 0.8$ eV has been determined [31]. This insensitivity of band offset on absorber surface stoichiometry corresponds to a significant deviation from the transitivity rule. Although the transitivity rule does not necessarily need to be fulfilled, its violation is in contradiction to the behavior of most semiconductor interfaces and in particular also to the observed transitivity at the interfaces with different II–VI compounds.

Since the influence of Cu-poor surfaces on the band alignment at the CIS/buffer layer interface remains uncertain, it is at present not possible to finally conclude whether band alignment is responsible for the saturation of the photovoltage of CIS solar cells with increasing bandgaps. From the experimental VBOs presented here, it is suggested that there is a cliff at the CuGaSe$_2$/CdS interface, which prevents the Fermi level from shifting close to the conduction band. However, this cliff is only about 0.2 eV if stoichiometric CuGaSe$_2$ is assumed (see Fig. 11.8), which might not be a severe limitation. But the cliff will become considerably larger in the case of a Cu-poor CuGaSe$_2$ surface if the band alignment as drawn in the right part of Fig. 11.9 is correct, since it is the difference between the conduction bands of CuGaSe$_2$ and CdS which is important, and not that between the 135 phase and CdS.

Another open question concerns the influence of the deposition technique on the band alignment. There are also not yet enough data to figure out if the chemical bath deposition (CBD) of CdS leads to a different band alignment. Neglecting effects of bandgap changes [31], it seems that the VBOs at CIS/CdS are not significantly different for evaporated (physical vapor deposition, PVD) and CBD CdS, as indicated by the results of Morkel et al. and by the identical VBO determined for PVD and CBD CdS on CuGaSe$_2$ [10, 46]. However, for CuInS$_2$/CdS widely different band alignments have been observed for PVD [4] and CBD CdS [47].

An alternative explanation for the saturation of photovoltage with increasing CIS bandgap is also related to interface formation, in particular to the self-compensation mechanism described in Sect. 11.4. However, only a few experiments on these effects are available so far. It can be expected that the induced defects not only limit band bending and thereby the photovoltages and conversion efficiencies, but also affect interface induced minority carrier recombination and chemical interface reactivity. So far, it is not generally clarified under which conditions defect formation occurs and how it affects the device behavior. Defect formation should also be affected by the buffer layer materials and its deposition technique, and also change with the substrate material as $CuInSe_2$, $CuGaSe_2$, and $CuInS_2$. In particular, we would expect that the covalency of the chemical bond in the different chalcopyrites has a strong influence on the defect formation.

Defect formation via self-compensation at interfaces has a fundamental difference to self-compensation in the bulk. The reason is that in bulk material all properties as Fermi-level position, chemical potentials and others are constant because of symmetry. For example if a Cu vacancy is created in bulk CIS by moving the Fermi level close to the conduction band, in the same instant an interstitial Cu is created. The complete defect pair is electrically neutral and will therefore not lead to self-compensation. To have only a compensating acceptor, the Cu must be removed from the lattice. But, in an isotropic material, the Cu_i concentration is constant and there is thus no driving force for diffusion. In contrast, if Cu is mobile and the Fermi-level position in the bandgap is not constant as at an interface, the interstitial Cu can diffuse to or from the interface (or grain boundary).

If the movement of the Fermi level during formation of the CIS/buffer layer interface leads to removal of Cu from the CIS surface and creates Cu vacancies, the effect might contribute to interdiffusion. It can be expected that the presence of (negatively charged) vacancies promotes interdiffusion of (positively charged) ions (Cd^{2+}), not only because of the attractive charge interaction but also because unoccupied lattice sites become available.

In summary, the experiments performed so far have identified different phenomena at the CIS/CdS interface. Unfortunately, there is not enough systematics as yet, in the experiments to conclude about the important issues of interfaces for the solar cell performance. It is not always possible to compare melt-grown single crystal surfaces and coevaporated thin film material. Furthermore the measurements on single crystals were done some years ago mainly using low excitation energy synchrotron radiation, which gave no access to deeply bound core-levels. Comprehensive data on conduction band alignment and interdiffusion at the CIS/buffer layer interface and their dependence on absorber and interface preparation are also missing. In the future, much more effort is necessary to better understand the CIS interface behavior, which undoubtedly is considerably more complex than those of conventional elemental or compound III–V and II–VI semiconductors. More systematic studies are required to understand the mechanisms of interface chemistry, band alignment and defect formation, and their dependence on preparation and processing steps.

Acknowledgments

An important contribution to the results described in this chapter is the preparation of the substrates, in particular of the CIS thin films with the Se caps. We are grateful to the collaboration with ZSW and particularly wish to thank R. Kniese and M. Powalla for the continuity and quality of this work. We also wish to thank W. Jaegermann, T. Löher, and R. Hunger for their contributions and fruitful discussions. Part of this work has been supported by the Bundesministerium für Bildung & Forschung (BMBF) in the framework of "Hochspannunsgnetz" (grant 01SF0114).

References

1. Ramanathan K., Contreras, M.A., Perkins, C.L., Asher, S., Hasoon, F.S., Keane, J., Young, D., Romero, M., Metzger, W., Noufi, R., Ward, J., Duda, A.: Properties of 19.2% efficiency ZnO/CdS/CuInGaSe$_2$ thin-film solar cells. Prog. Photovolt.: Res. Appl. 11, 225–230 (2003)
2. Contreras, M.A., Nakada, T., Hongo, M., Pudov, A.O., Sites, J.R.: ZnO/ZnS(O,OH)/Cu(In,Ga)Se$_2$/Mo solar cell with 18.6% efficiency. In Proceedings of the 3rd World Conference on Photovoltaic Energy Conversion, Osaka, vol. 1, 570–573 (2003)
3. Hüfner, S.: Photoelectron Spectroscopy. Springer, Berlin Heidelberg New York (1995)
4. Klein, A., Löher, T., Pettenkofer, C., Jaegermann, W.: Band lineup between CdS and ultra high vacuum-cleaved CuInS$_2$ single crystals. Appl. Phys. Lett. 70, 1299 (1997)
5. Löher, T., Jaegermann, W., Pettenkofer, C.: Formation and electronic properties of the CdS/CuInSe$_2$ (0 1 1) heterointerface studied by synchrotron-induced photoemission. J. Appl. Phys. 77, 731 (1995)
6. Mönch, W.: Semiconductor Surfaces and Interfaces. Springer, Berlin Heidelberg New York (1993)
7. Franciosi, A., Van de Walle, C.G.: Heterojunction band offset engineering. Surf. Sci. Rep. 25, 1 (1996)
8. Yu, E.T., McCaldin, J.O., McGill, T.C.: Band offsets in semiconductor heterojunctions. In: Ehrenreich, H., Turnbull, D. (eds.) Solid State Physics, vol. 46. Academic Press, Boston (1992)
9. Zhang, S.B., Wei, S.-H., Zunger, A.: A phenomenological model for systemization and prediction of doping limits in II–VI and I–III–VI$_2$ compounds. J. Appl. Phys. 83, 3192 (1998)
10. Klein, A., Fritsche, J., Jaegermann, W., Schön, J.H., Kloc, C., Bucher, E.: Fermi level dependent defect formation at Cu(In,Ga)Se$_2$ interfaces. Appl. Surf. Sci. 166, 508 (2000)
11. Bauknecht, A., Blieske, U., Kampschulte, T., Albert, J., Sehnert, H., Lux-Steiner, M.C., Klein, A., Jaegermann, W.: Band offsets at the ZnSe/CuGaSe$_2$(001) heterointerface. Appl. Phys. Lett. 74, 1099 (1999)
12. Nelson, A.J., Niles, D.W., Schwerdtfeger, C.R., Wei, S.-H., Zunger, A., Höchst, H.: Prediction and observation of II–VI/CuInSe$_2$ heterojunction band offsets. J. Electron Spectrosc. Relat. Phenom. 68, 185–193 (1994)

13. Nelson, A.J., Schwerdtfeger, C.R., Wei, S.-H., Zunger, A., Rioux, D., Patel, R., Höchst, H.: Theoretical and experimental studies of the ZnSe/CuInSe$_2$ heterojunction band offset. Appl. Phys. Lett. 62, 2557 (1993)
14. Wei, S.-H., Zunger, A.: Band offsets and optical bowings of chalcopyrites and Zn-based II–VI alloys. J. Appl. Phys. 78, 3846 (1995)
15. Schulmeyer, T., Hunger, R., Klein, A., Jaegermann, W., Niki, S.: Photoemission study and band alignment of the CdS/CuInSe$_2$(001) heterojunction. Appl. Phys. Lett. 84, 3067 (2004)
16. Hunger, R., Scheer, R., Diesner, K., Su, D., Lewerenz, H.J.: Heteroepitaxy of CuInS$_2$ on Si(1 1 1). Appl. Phys. Lett. 69, 3010 (1996)
17. Niki, S., Makita, Y., Yamada, A., Hellman, O., Fons, P.J., Obara, A., Okada, Y., Shioda, R., Oyanagi, H., Kurafuji, T., Chichibu, S., Nakanishi, H.: Heteroepitaxy and characterization of CuInSe$_2$ on GaAs(001). J. Cryst. Growth 150, 1201–1205 (1995)
18. Metzner, H., Hahn, T., Cieslak, J., Grossner, U., Reislohner, U., Witthuhn, W., Goldhahn, R., Eberhardt, J., Gobsch, G., Krausslich, J.: Epitaxial growth of CuGaS$_2$ on Si(111). Appl. Phys. Lett. 81, 156–158 (2002)
19. Bauknecht, A., Siebentritt, S., Gerhard, A., Harneit, W., Brehme, S., Albert, J., Rushworth, S., Lux-Steiner, M.C.: Defects in CuGaSe$_2$ thin films grown by MOCVD. Thin Solid Films 361–362, 426 (2000)
20. Turcu, M., Kötschau, I.M., Rau, U.: Composition dependence of defect energies and band alignments in the Cu(In$_{1-x}$Ga$_x$)(Se$_{1-y}$S$_y$)$_2$ alloy system. J. Appl. Phys. 91, 1391 (2002)
21. Caldas, M.J., Fazzio, A., Zunger, A.: A universal trend in the binding energies of deep impurities in semiconductors. Appl. Phys. Lett. 45, 671 (1984)
22. Zunger, A.: Composition dependence of deep impurity levels in alloys. Phys. Rev. Lett. 54, 849 (1985)
23. Langer, J.M., Delerue, C., Lannoo, M., Heinrich, H.: Transition-metal impurities in semiconductors and heterojunction band lineups. Phys. Rev. B 38, 7723 (1988)
24. Schmid, D., Ruckh, M., Schock, H.W.: Photoemission studies on Cu(In,Ga)Se$_2$ thin films and related binary selenides. Appl. Surf. Sci. 103, 409 (1996)
25. Schulmeyer, T., Hunger, R., Jaegermann, W., Klein, A., Kniese, R., Powalla, M.: Interface formation between polycrystalline Cu(In,Ga)Se$_2$ and II–VI-compounds. In: Proceedings of the 3rd World Conference on Photovoltaic Energy Conversion, Osaka, vol. 1, pp. 364–367 (2003)
26. Schulmeyer, T., Kniese, R., Hunger, R., Jaegermann, W., Powalla, M., Klein, A.: Influence of Cu(In,Ga)Se$_2$ band gap on the valence band offset with CdS. Thin Solid Films 451–452, 420 (2004)
27. Zhang, S.B., Wei, S.-H., Zunger, A., Katayama-Yoshida, H.: Defect physics of the CuInSe$_2$ chalcopyrite semiconductor. Phys. Rev. B 57, 9642 (1998)
28. Fritsche, J., Schulmeyer, T., Kraft, D., Thißen, A., Klein, A., Jaegermann, W.: Utilization of sputter depth profiling for the determination of band alignment at polycrystalline CdTe/CdS heterointerfaces. Appl. Phys. Lett. 81, 2297 (2002)
29. Shafarman, W.N., Klenk, R., McCandless, B.E. Device and material characterization of Cu(InGa)Se$_2$ solar cells with increasing band gap. J. Appl. Phys. 79, 7324 (1996)
30. Herberholz, R., Nadenau, V., Rühle, U., Köble, C., Schock, H.W., Dimmler, B. Prospects of wide-gap chalcopyrites for thin film photovoltaic modules. Sol. Energy Mater. Sol. Cells 49, 227 (1997)

31. Morkel, M., Weinhardt, L., Lohmüller, B., Heske, C., Umbach, E., Riedl, W., Zweigart, S., Karg, F.: Flat conduction-band alignment at the CdS/CuInSe$_2$ thin-film solar-cell heterojunction. Appl. Phys. Lett. 79, 4482 (2001)
32. Löher, T., Klein, A., Pettenkofer, C., Jaegermann, W.: Partial density of states in the CuInSe$_2$ valence bands. J. Appl. Phys. 81, 7806, (1997)
33. Jaffe, J.E., Zunger, A.: Anion displacements and the band-gap anomaly in ternary ABC$_2$ chalcopyrite semiconductors. Phys. Rev. B 27, 5176 (1983)
34. Liao, D., Rockett, A.: Cu depletion at the CuInSe$_2$ surfaces. Appl. Phys. Lett. 82, 2829 (2003)
35. Tiwari, A.N., Blunier, S., Filzmoser, M., Zogg, H., Schmid, D., Schock, H.W. Characterization of heteroepitaxial CuIn$_3$Se$_5$ and CuInSe$_2$ layers on Si substrates. Appl. Phys. Lett. 65, 3347 (1994)
36. Gartsman, K., Chernyak, L., Lyahovitskaya, V., Cahen, D., Didik, V., Kozlovsky, V., Malkovich, R., Skoryatina, E., Usacheva, V.: Direct evidence for diffusion and electromigration of Cu in CuInSe$_2$. J. Appl. Phys. 82, 4282 (1997)
37. Turcu, M., Pakma, O., Rau, U.: Interdependence of absorber composition and recombination mechanism in Cu(In,Ga)(Se,S)$_2$ heterojunction solar cells. Appl. Phys. Lett. 80, 2598 (2002)
38. Rau, U., Schock, H.W.: Electronic properties of Cu(In,Ga)Se$_2$ heterojunction solar cells—recent achievements, current understanding, and future challenges. Appl. Phys. A 69, 131 (1999)
39. Zhang, S.B., Wei, S.-H., Zunger, A.: Stabilization of ternary compounds via ordered arrays of defect pairs. Phys. Rev. Lett. 78, 4059 (1997)
40. Klein, A., Jaegermann, W.: Fermi-level dependent defect formation in Cu-chalcopyrite semiconductors. Appl. Phys. Lett. 74, 2283 (1999)
41. Klein, A., Löher, T., Pettenkofer, C., Jaegermann, W.: Chemical interaction of Na with cleaved (0 1 1) surfaces of CuInSe$_2$. J. Appl. Phys. 80, 5039 (1996)
42. Chadi, D.J.: Predictor of p-type doping in II–VI semiconductors. Phys. Rev. B 59, 15181 (1999)
43. Walukiewicz, W.: Intrinsic limitations to the doping of wide-gap semiconductors. Physica B 302–303, 123 (2001)
44. Weast, R.C.: CRC Handbook of Chemistry and Physics. CRC Press, Boca Raton (1985)
45. Trasatti, S.: The absolute electrode potential: an explanatory note (Recommendations 1986). J. Electroanal. Chem. 209, 417 (1986)
46. Nadenau, V., Braunger, D., Hariskos, D., Schock, H.W.: Characterization and optimization of CuGaSe$_2$/CdS/ZnO heterojunctions. Inst. Phys. Conf. Ser. 152, 955 (1998)
47. Hashimoto, Y., Takeuchi, K., Ito, K.: Band alignment at CdS/CuInS$_2$ heterojunction. Appl. Phys. Lett. 67, 980 (1995)

Bandgap Variations for Large Area Cu(In,Ga)Se$_2$ Module Production

R. Kniese, G. Voorwinden, R. Menner, U. Stein, M. Powalla, and U. Rau

The development of large area Cu(In,Ga)Se$_2$ photovoltaic (PV) modules has reached a remarkable level. Nevertheless, scope for further improvement lies in the ability of the Cu(In,Ga)(Se,S)$_2$ alloy system to vary the bandgap via the In-to-Ga or the S-to-Se ratio. This chapter investigates the possible gain in efficiency resulting from a better match to the solar spectrum, reduced losses related to the monolithic integration and an improved temperature coefficient, all of which are achievable by increasing the bandgap energy of the absorber material. Experimental results are compared to theoretical predictions and the deviations are discussed. Furthermore, graded bandgap structures, obtainable by introducing compositional profiles in the depth of the absorber, have demonstrated their potential to improve the conversion efficiency. Nevertheless, so far the potential of graded bandgap structures has not been fully exploited for large area module production. In this chapter, the technical feasibility of depositing absorbers with a graded bandgap on a large area is shown and the effect on the electrical characteristic is discussed.

12.1 Introduction

Thin-film Cu(In,Ga)(Se,S)$_2$ is a promising material for the production of highly efficient and cost-saving PV modules. Conversion efficiencies up to 19.2% for small area laboratory cells [1] have been demonstrated. The key issue to succeed in the production of low-cost PV modules is the deposition of laterally homogenous Cu(In,Ga)(Se,S)$_2$ absorber material on a large area.

Two kinds of processes have proven their suitability for large-scale Cu(In,Ga)(Se,S)$_2$ deposition. In the so-called *sequential process*, the absorber is produced by selenisation of metal precursors. The precursors are obtained from sequential sputtering of the individual elements. The final composition and homogeneity depends on the precision of these sputter processes. Efficiencies of 13.1% have been achieved on 4,937 cm^2 [2].

The second approach is the so-called *in-line co-evaporation process*. In this process, the elements Cu, In, Ga and Se are evaporated simultaneously in one process step on continuously moving substrates. Efficiencies of 12.2% have been achieved on 6,507 cm^2 [3] with this process.

The lower module efficiencies compared to those of small area laboratory cells have manifold reasons which must be considered when aiming for high module efficiencies. Inhomogeneities of the absorber material can lead to variations in the current–voltage characteristics of the cells connected in series, forcing each of the cells to operate away from its maximum power point. Minimization of this effect requires excellent homogeneity of the absorber material over the entire module area.

Further losses that arise from the monolithic integration involve: enhanced absorption in the window layer, due to its higher thickness, needed because lower resistivity is needed; area losses due to the interconnects of the individual cells; and series resistance. One opportunity to minimize these losses lies in the ability to increase the bandgap energy E_g of $Cu(In_{1-x}Ga_x)(Se_{1-y}S_y)_2$ via the Ga/(Ga + In) ratio x or the S/(S + Se) ratio y. The short-circuit current density j_{sc} is reduced thereby, reducing series resistance losses, and the open-circuit voltage V_{OC} per cell is increased.

Besides reducing the series resistance related losses, a higher bandgap would lead to an improved match to the solar spectrum and a reduced efficiency decline at high temperatures. Unfortunately, all attempts to increase the bandgap energy above its standard value of about 1.2 eV have so far reduced the efficiency. In Sect. 12.3, experimentally obtained solar cell output parameters for different bandgap energies are compared to theoretically achievable values, and the losses leading to the deviations are discussed.

The effect of an increased bandgap energy on the module efficiency is quantified by using the numerical device simulator SCAPS developed by Burgelmann and coworkers [4]. Assumptions must be made for the module simulation regarding the dependence of the solar cell parameters on the bandgap energy. To discuss these dependencies, experimental results are compared to numerical calculations on the basis of the one-diode model with two different interdependencies of the ideality factor A on the bandgap energy.

The ability to change the bandgap energy of $Cu(In_{1-x}Ga_x)Se_2$ allows the introduction of graded band edges via compositional profiles. The to-date most efficient solar cells are prepared with the three-stage process [5] involving an increasing conduction band edge towards the back-contact. Apart from a better absorber quality [6], the superiority of this process is ascribed to the graded conduction band edge [7]. The technical feasibility of depositing absorber material with controlled bandgap grading on a large area with in-line co-evaporation is demonstrated and the effect on the solar cell parameters is discussed in Sect. 12.4.

12.2 Device Preparation and Characterisation

The $Cu(In,Ga)Se_2$ films are deposited by co-evaporation of the elements copper, indium, gallium, and selenium from separate sources onto moving substrates. The modules and cells are prepared in a deposition chamber that is capable of processing $30 \times 30\,\text{cm}^2$ substrates (see Fig. 12.1). The arrangement of the evaporation sources is schematically illustrated in Fig. 12.2. Due to the line-shaped evaporation profile of the sources, the films are homogenously perpendicular to the transport axis. The copper, indium, and gallium evaporation rates are controlled by atomic absorption spectroscopy [8] coupled to the chamber with optical fibres. The selenium flux is a few times the amount necessary for the formation of Cu(In,Ga)Se₂. We use soda-lime glass coated with DC-sputtered molybdenum as a substrate material. The Cu(In,Ga)Se₂ films grow to a thickness of $2\,\mu\text{m}$ within 15 min. The substrate temperature is fairly constant around 600°C during the entire film growth. The composition and thickness of the Cu(In,Ga)Se₂ films are measured by X-ray fluorescence analysis (XRF). The information depth of XRF is several times the layer thickness and thus gives good averaged values through the bulk of the film. Depth profiles are provided by sputtered neutral mass spectrometry (SNMS).

Solar cells are fabricated by depositing a CdS buffer layer by a chemical bath, RF-sputtered i-ZnO, and DC-sputtered n-ZnO:Al. Ni/Al grids serve as the front contact. The cells have an area of $0.5\,\text{cm}^2$. The devices are then characterized by quantum efficiency (QE) measurements using monochromatic chopped light without background illumination. From these measurements, the bandgaps were determined by extrapolating $(h\nu \cdot \text{QE})^2$ vs photonenergy $h\nu$. This assumes that QE is proportional to the absorption coefficient. The current–voltage measurements were performed at 25°C under a simulated AM1.5 spectrum. The light intensity was adjusted in each case based on QE data.

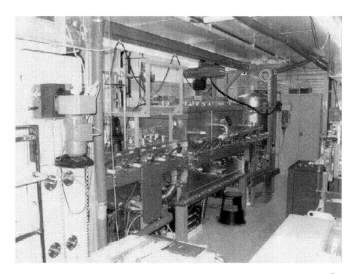

Fig. 12.1. In-line deposition chamber capable of processing $30 \times 30\,\text{cm}^2$ substrates

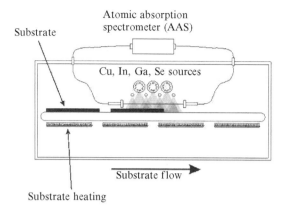

Fig. 12.2. Schematic view of the evaporation source arrangement and substrate transport

12.3 Cu(In$_{1-x}$Ga$_x$)Se$_2$ with Varying Bandgap: Theoretical Potential and Experimental Results

12.3.1 Solar Cell Efficiency from the One-Diode Model

This section provides a phenomenological description of the development of PV output parameters of Cu(In,Ga)Se$_2$ solar cells with increasing Ga content in terms of a simple one-diode model. The results of these calculations are compared to a series of experimental solar cells with various Ga contents. For the model, we assume that the current density j vs voltage U characteristic of the solar cell is given by the diode equation [9]

$$j = j_0 \left(e^{e(U - jR_S)/AkT} - 1 \right) + j_{ph} + \frac{U - jR_S}{R_P}, \tag{12.1}$$

with the thermal voltage kT/e, the diode ideality factor A, the series and parallel resistance R_S and R_P, respectively, and the photogenerated current density j_{ph}. Note that the short-circuit current density $j_{sc} = j(U = 0) \approx j_{ph}$ for sufficiently high R_S. The saturation current density j_0 is given by

$$j_0 = j_{00}\, e^{-Ea/AkT}, \tag{12.2}$$

where j_{00} is a weakly temperature dependent prefactor. The influence of the different recombination mechanisms occurring in Cu(In,Ga)Se$_2$ solar cells can be accounted for by the use of the appropriate diode ideality factor A and activation energy E_a. Hence, the structure of Eqs. 12.1 and 12.2 is not affected by the recombination mechanism [9].

To calculate the dependence of the short-circuit current density j_{sc}, the open-circuit voltage V_{OC}, the fill factor FF and the efficiency η on the bandgapenergy from Eq. 12.1, assumptions regarding the interdependence between the diode parameters and the bandgap energy E_g are necessary.

The photocurrent j_{sc} is related to the solar spectrum and E_g by

$$ j_{SC} = Q\, e \int\limits_{h\nu = Eg}^{\infty} \frac{dn_{Ph}}{dh\nu}\, d\,(h\nu), \qquad (12.3) $$

with $dn_{Ph}/d(h\nu)$ being the photon flux per energy interval under AM1.5 conditions and Q a prefactor that accounts for reflectance and absorption in the window layer and imperfect carrier collection. A QE of one is assumed.

The value of j_0 is derived from (12.2) If for different bandgap energies there were no changes in the absorber quality or recombination mechanism, j_{00} and A would remain constant. Experimentally, an increase of A is observed with increasing E_g, e.g., by increasing the Ga content. This observation is explained by an increasing tunneling contribution to the recombination [10]. To demonstrate the consequence of this effect, an additional calculation was performed in which the diode ideality factor was increased linearly from 1.3 for $x = 0.1$ to 1.7 for $x = 0.76$. The values assumed for the calculation are compared to the experimental values in the inset of Fig. 12.3(a).

The activation energy E_a equals E_g in the case of recombination in the space charge region [9]. Experimental results from different laboratories confirm the validity of this assumption throughout the composition range from pure CuInSe$_2$ to pure CuGaSe$_2$ in the case of Cu-poor compositions [10,12,13]. The parameters Q and j_{00} were adjusted to typical values of standard cells with $E_g = 1.16\,\mathrm{eV}$, leading to an active-area efficiency of 16.2%. For the different bandgap energies V_{OC} is calculated from Eqs. 12.1 and 12.2, the FF from the standard phenomenological expression [14] and j_{sc} from the integral Eq. 12.3.

Figure 12.3(a) and (b) depicts η, V_{OC}, FF, and j_{sc} calculated as a function of the bandgap energy for a constant ideality factor and for one increasing with increasing bandgap energy. In the case of a constant ideality factor A, the theoretical benefit of an increasing bandgap is obvious. The conversion efficiency increases and reaches its maximum value of 18% at a bandgap of about 1.4 eV. Furthermore, the FF increases continuously with increasing bandgap energy. When assuming an increasing ideality factor, we find that the increase of V_{OC} and FF with increasing x is considerably reduced. This assumption would lead to an almost constant efficiency within a range $1\,\mathrm{eV} \le E_g \le 1.4\,\mathrm{eV}$.

In contrast to the calculations, the experimental data in Fig. 12.3(a) indicate a drop of the efficiencies if the Ga content is either increased or decreased from the standard value of $x = 0.3$. The experimental open-circuit voltages are below the calculated values. This is due to the fact that for the calculation j_{00} was assumed to be constant. Note that j_{00} depends on the minority carrier lifetime and therefore reflects the absorber quality. Thus, we conclude that

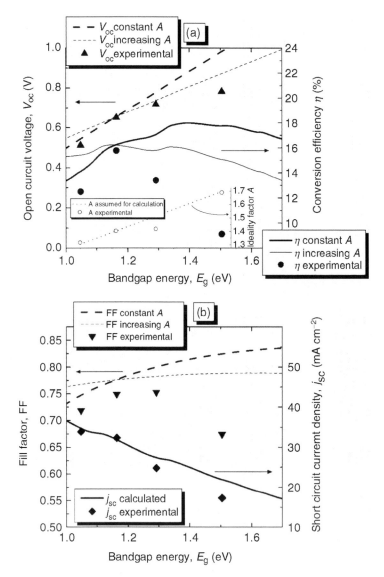

Fig. 12.3. (a) Efficiency η and open-circuit voltage V_{OC}; (b) Short-circuit current density j_{SC} and fill factor FF calculated from the one-diode model with constant or increasing ideality factor A, compared to experimental results

the low experimental open-circuit voltages for bandgap energies $E_g \neq 1.15\,\text{eV}$ result from a poorer absorber quality for Ga contents x that are either higher or lower than the optimum value [15]. Up to now, it is not yet clear whether this poorer absorber quality is a general property of the material or is due to the fact that our process is optimized for the standard value of $x = 0.3$.

From comparing the experimental data with the calculated curves for the FF in Fig. 12.3(b), we recognize that, besides the low V_{OC}, the efficiency loss results also from an FF that is much lower than expected in the case of high Ga content. Additionally, we observe that the experimental j_{sc} values in Fig. 12.3(b) are somewhat lower than the calculated ones. These losses in FF and j_{sc} in the case of the solar cells with high Ga contents result from the small diffusion length in these absorber layers leading to a poor charge carrier collection that even becomes voltage dependent [16]. Figure 12.4 compares the QE of cells with Ga contents $x = 0.28$ and $x = 0.76$. The loss in j_{sc} for the high-Ga sample occurs in the long wavelength range, caused by a short collection length. Also shown in Fig. 12.4 are the relative changes $QE(-0.8V)/QE(0V) - 1$ as a measure for the voltage bias dependence of carrier collection. For the sample with $x = 0.28$, the relative difference between the QE measured at a voltage $V = -0.8$ and that at zero bias is below 1% almost in the entire wavelength range. In contrast, for the sample with $x = 0.76$, the same quantity exhibits a continuous increase up to 5% with increasing wavelength. Such a spectral pattern is expected if the voltage dependence of the photocurrent results from a short diffusion length.

12.3.2 Temperature Dependence of Module Performance

The temperature T of a PV module under operating conditions is higher than the temperature $T = 25°C$ defined by the standard test conditions (STC). A standard temperature for different module types is given by their normal operating cell temperature (NOCT). This is the temperature that a module reaches under an illumination intensity of 800 W/m², an ambient temperature of 20°C and a wind velocity of 1 m/s. Common NOCT values for Cu(In,Ga)Se₂ modules are close to 50°C. However, maximum operating temperatures of up to 80°C can be reached under practical outdoor conditions. Therefore, it is

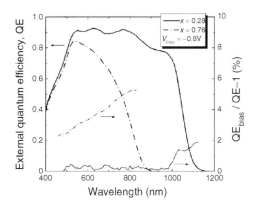

Fig. 12.4. External quantum efficiency QE and relative change of QE under reverse bias voltage $V_{Bias} = -0.8\,V$ of Cu(In$_{1-x}$Ga$_x$)Se₂ solar cells with $x = 0.28$ and 0.76

obvious that the conversion efficiency η at elevated temperatures has much more practical significance than the efficiency under STC.

The temperature dependence of η is primarily a result of the decline of the open-circuit voltage V_{OC} and the FF with increasing temperature. The output parameters η, V_{OC}, and FF can be calculated from Eq. 12.1 and 12.2 with the parameters summarised in Table 12.1. The temperature dependence of j_{00} depends on the recombination mechanism [9]. Since this temperature dependence is weak, it has been neglected in the following calculations.

The output parameters η, V_{OC}, and FF calculated from Eq. 12.1 and 12.2 for different bandgap energies are normalized to their value at 300 K and plotted vs temperature in Fig. 12.5. An increased bandgap energy obviously leads to a reduced relative decline of all output parameters with increasing temperature. To estimate the influence of the module temperature on the annual energy production, the mean temperature of a module can be approximated by the NOCT value, which is around 325 K.

Table 12.1. Diode parameters for the cell efficiency calculation

A	j_{00}	R_S	R_P	Q
1.4	$39{,}000\,\mathrm{A/cm^2}$	$0.25\ \Omega\,\mathrm{cm^2}$	$2\,\mathrm{k\Omega\,cm^2}$	0.77

A diode ideality factor, j_{00} saturation current prefactor, R_S series resistance, R_P parallel resistance, Q photocurrent prefactor.

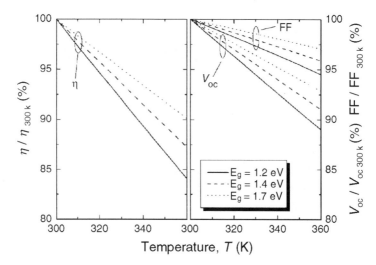

Fig. 12.5. Temperature dependence of the conversion efficiency η, open-circuit voltage V_{OC}, and fill factor FF normalized to the values at 300 K for different bandgap energies calculated from Eqs. 12.1 and 12.2

12.3.3 Module Simulation

The influence of the bandgap energy of the absorber material on the module efficiency was simulated with a software developed by Burgelman et al. [4]. This software allows for the calculation of the efficiency of a monolithically integrated module from the diode parameters of the underlying cell. The results from Sect. 12.3.1 for the different bandgap energies were used as input parameters. The values resulting from constant A and j_{00} were used to obtain the module efficiency for the case that there is no change in the recombination mechanism or the absorber quality with increasing bandgap. The cell parameters are summarized in Table 12.2. The geometry of the interconnects is illustrated in Fig. 12.6 and the corresponding parameters are summarised in Table 12.3. Looking at the quantity $g1$ in Fig. 12.6, it can be seen that the interconnection via the absorber layer creates a shunt path for each cell, which depends on the absorber conductivity. The detrimental effect of this shunt is more significant for higher bandgap materials with a respectively higher voltage at working conditions. In order to determine the most optimistic potential of the high bandgap material, this effect was eliminated by assuming an unrealistically low conductivity of the absorber material. In future production, this effect could be minimized by overlapping of $g1$ and $g2$. The interdependence between the sheet resistance R_{sq} and the transmission T is described by

$$T = T_2 - (R_2/R_{sq})^m \tag{12.4}$$

with the parameters T_2, R_2, and m as summarised in Table 12.3 [4].

Table 12.2. Cell parameters for module simulation

E_g (eV)	1.1	1.2	1.4	1.7	A	1.4
V_{OC} (mV)	594	691	884	1,170	R_P [kΩcm^2]	2
j_{sc} (mA/cm^2)	33.4	30.3	24.9	17	R_S [Ωcm^2]	0.25
η (%)	15.2	16.4	18	16.6	T_{ZnO}	0.96
FF	76.2	78.5	81.6	83.6		

V_{OC} open-circuit voltage, j_{sc} short-circuit current density, η conversion efficiency, FF fill factor, A ideality factor, R_P and R_S parallel and series resistances, T_{ZnO} transmission of cell ZnO

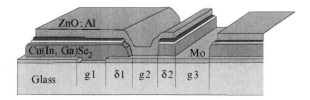

Fig. 12.6. Schematic illustration of the monolithic integration

Table 12.3. Parameters for module geometry and ZnO model

interconnect		ZnO model (12.4)	
$g1, g2, g3$	$50\,\mu m$	$T2$	0.96
$\delta1, \delta2$	$150\,\mu m$	$R2$	$3.3\,\Omega$
R_C	$2 \times 10^{-3}\,\Omega\,cm^2$	M	3.3

$g1, g2, g3, \delta1, \delta2$ parameters describing geometry of interconnect (cf. Fig. 12.6), R_C ZnO–Mo contact resistance, $T2,R2,m$ parameters for ZnO model Eq. 12.4

Increasing the bandgap energy leads to a higher open-circuit voltage per cell and a lower short-circuit current density. A higher bandgap therefore reduces losses when connecting cells to a module. The difference in these losses for different bandgap energies can be demonstrated by comparing the ratios $\eta_{module}/\eta_{cell}$ between the cell and the module efficiency at different bandgap energies.

An increased ZnO thickness reduces losses due to the series resistance, but increases the absorption in the window layer. As a consequence, there is an optimum ZnO thickness for every cell width. We first simulate the modules with given cell widths in dependence of the ZnO thickness and the corresponding sheet resistance. We then plot the results obtained for an optimised ZnO thickness vs the cell width (Fig. 12.7a). In this representation, a narrower cell implies more area losses due to the interconnects of the individual cells, thus a reduced effective PV active area of the module. In contrast, with increasing cell width the FF declines and the higher cell current requires a thicker ZnO, reducing its transmission. Hence, for every E_g there exists an optimum cell width. Figure 12.7a shows that an increase of E_g leads to a reduced loss in FF and transmission when comparing modules with the same design. The optimal cell width as well as the achievable ratio $\eta_{module}/\eta_{cell}$ increase for increasing bandgap energy.

For many applications, however, the dimensions as well as output voltage of the module are predefined. Consider, as an example, the voltage window of a DC–AC inverter. If we fix the width W_{module} of a module, the width $w_{cell} = W_{module}/n$ is given by the number n of the cells in the module. If we fix at the same time the open-circuit voltage $V_{OC,mod}$ of the module, the cells of the module have to fulfill

$$V_{OC,cell}/w_{cell} = V_{OC,mod}/W_{module}. \tag{12.5}$$

Because of Eq. 12.5, it is instructive to replot the data of Fig. 12.7a in order to compare modules with the same ratio $V_{OC,cell}/w_{cell}$, i.e., the same open-circuit voltage per cell width (Fig. 12.7b). First of all, we see from Fig. 12.7b that for a given $V_{OC,cell}/w_{cell}$, the ratio $\eta_{module}/\eta_{cell}$ is *always* higher for the high bandgap materials. In this case, the beneficial effect of the higher open-circuit voltage mainly is due to the reduced area loss due to fewer

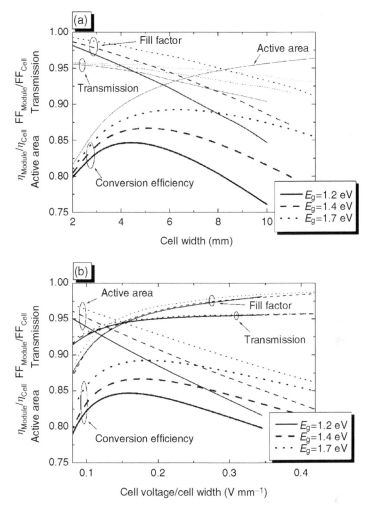

Fig. 12.7. Transmission, active area and ratio between module and cell efficiencies and their FF for different bandgap energies vs (**a**) the cell width and (**b**) open-circuit voltage per cell divided by the cell width

interconnects. Furthermore, the maximum in the efficiency curve is most shallow for the wide-gap material, thus allowing the highest flexibility for a module design.

The absolute module efficiencies are shown in Fig. 12.8. It can be seen that the efficiency gain due to better match to the solar spectrum of the 1.4 eV module is higher than the efficiency gain for the 1.7 eV module due to lower losses from cell interconnection. Experimental results from two batches of ten mini modules each, with an aperture area of 48.2 cm² with different bandgaps, are shown in Table 12.4. Unfortunately, the efficiency gain predicted under the

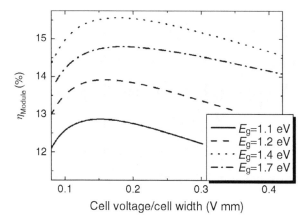

Fig. 12.8. Module efficiencies calculated from the cell parameters summarised in Table 12.2

Table 12.4. Photovoltaic output parameters of $10 \times 10\,\text{cm}^2$ modules, with an aperture area of $48.2\,\text{cm}^2$ and with different bandgap energies

	η (%)	V_{OC} (mV)/cell	FF (%)	j_{SC} (mA/cm^2)
$x = 0.32$,				
$E_{\text{g}} = 1.13\,\text{eV}$				
Average	11.1	612.3	62.0	29.0
Minimum	9.6	578.2	57.3	28.3
Maximum	12.8	646.5	66.7	30.0
$x = 0.6$,				
$E_{\text{g}} = 1.37\,\text{eV}$				
Average	10.4	777.1	67.0	19.9
Minimum	9.7	758.5	65.1	19.3
Maximum	11.1	791.3	69.2	20.4

optimistic assumptions leading to Fig. 12.8 is not achieved by increasing the bandgap energy in the real modules due to additional losses already discussed in Sect. 12.3.1.

12.4 Bandgap Grading

12.4.1 Introduction

The CdS–Cu(In,Ga)(Se,S)$_2$–Mo heterostructure contains regions of high recombination that may impede high PV conversion efficiencies. The CdS–Cu(In,Ga)(Se,S)$_2$ interface and the continuous metallic back-contact, both in the range of the minority carriers, which are holes in the first case and electrons in the second case, should cause substantial recombination. Fortunately, the formation of a Cu-deficient surface layer with a lowered valence band [17] and a MoSe$_2$ layer with a higher bandgap [18] reduces the concentration of the minority carriers at the corresponding interfaces and therefore facilitates the high conversion efficiencies already reached with this heterostructure. The effect resulting from graded band edges can also be achieved by appropriate band bending induced by positive interface charges at the CdS–Cu(In,Ga)(Se,S)$_2$ interface [19] and negative charges at the back-contact, realised in silicon solar cells by a highly p-doped layer [20].

The outstanding opportunity of the Cu(In,Ga)(Se,S)$_2$ alloy system is the ability to shift the conduction and the valence band edges by varying the In-to-Ga and the S-to-Se ratio, respectively [12]. The positive effects of the Cu-deficient surface layer and the MoSe$_2$ layer can also be obtained by additional compositional profiles with increasing Ga content towards the back-contact, the so-called "normal grading" [21], and increasing the S content towards the CdS–Cu(In,Ga)(Se,S)$_2$ interface [22]. As the band bending is likely defined by Fermi-level pinning at interface states [23], the dependence of the energetic position of these states on the interfacial composition has to be as close as possible to the conduction band in order to minimize recombination at the buffer/absorber interface as well as resistive losses for the electron transport across this interface [12, 24].

A further area of high recombination is the space charge region. Numerical simulations show that an increased bandgap in the space charge region can enhance the open-circuit voltage without significantly decreasing the short-circuit current density [25,26]. This suggests a so-called "double graded" structure with increasing conduction band edge towards the interface and towards the back-contact [27].

Besides the ability to screen areas of high recombination from minority carriers, graded band edges and doping profiles may be used to enhance carrier collection in low-lifetime materials [28]. In Sect. 12.3.1, it has been shown that devices with high Ga content suffer from poor carrier collection properties resulting from a short diffusion length. The introduction of an increasing conduction band edge towards the back-contact may improve the carrier collection properties in this material [29].

12.4.2 Deposition of Graded Absorbers

The preparation of graded absorbers in an in-line system presents a special challenge. Compositional profiles in the depth of the absorber can be obtained by the spatial variation of the elemental flux distributions along the transport direction. They result from the arrangement of the line sources. One example for element-specific rate distributions is illustrated in Fig. 12.9. The compositional profile impressed by the changing supply of the elements gets partly equalized by migration.

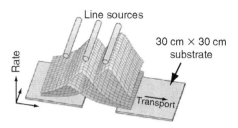

Fig. 12.9. Evaporation rate profiles of line sources

We prepared samples with four different gradients that we refer to as A, B, C, and D. For all samples, the Ga/(Ga + In) ratio increases towards the Mo back-contact with various slopes. The steepness of the gradient increases from sample A to D, whereas the integral Ga/(Ga + In) ratio of all samples is kept within the range of 0.29–0.32.

During processes A and B, the absorber composition is Cu-rich for the first 1 µm and 1.5 µm of layer growth, respectively, and can therefore be denoted as a bilayer process. We cannot realize bilayer processes for steeper gradients due to geometrical reasons. Processes C and D are inverted processes with only Cu-poor growth. All four samples have the same final copper concentration and a final thickness of 2 µm.

We numerically simulated the Ga/(Ga + In) gradient impressed by the evaporation rate profiles with a finite element method (FEM) [30]. These data are plotted as straight solid lines in the left column of Fig. 12.10a, assuming no intermixing in the films during growth. In case of C, we added measured condensation rates (open symbols) to the data [31]. The right column of Fig. 12.10a shows the corresponding SNMS depth profiles. We used XRF compositional data to quantify both the FEM and SNMS data. The simulated and measured Ga/(Ga + In) fit well at the absorber's surface [see also lower part of Fig. 12.10b], indicating a very low intermixing there. Towards the back-contact the SNMS curves are smeared out. A possible explanation is the limited depth resolution of SNMS as well as intermixing during growth.

12.4.3 Analysis of Solar Cells with Graded Absorbers

The solar cell output parameters of the devices prepared with processes A–D are summarised in Table 12.5. Moreover, V_{OC} and j_{sc} are depicted in Fig. 12.10b. The most striking issue is an increase of j_{sc} with increasing steepness of the Ga gradient. The reason for this finding is shown by the QE measurements displayed in Fig. 12.11a. The absorption edge shifts to longer wavelengths since the minimum bandgap decreases with increasing steepness of the Ga gradient. In samples prepared from graded absorbers, the absorption edge should equal the minimum E_g in the absorber, assuming this part is thick enough to provide for substantial absorption. In Fig. 12.11b, we compare the bandgap calculated [32] from the minimum Ga/(Ga + In) ratios found

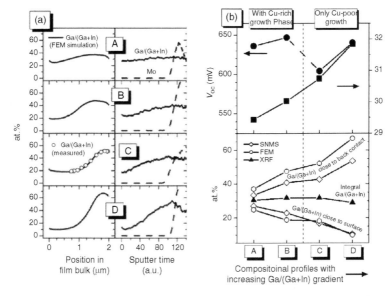

Fig. 12.10. (a) Ga gradients of processes A– to D from FEM simulations (*left row*) and SNMS measurements of the Ga/(Ga + In) -ratio and of Mo (*right row, full and dashed lines,* respectively);(b) Short-circuit current density j_{SC}, open-circuit voltage V_{OC}, and Ga content at the back-contact, the surface and integral for the cells for processes A–D

Table 12.5. Device parameters of Cu(In,Ga)Se₂ solar cells with increasing Ga grading from sample A–D

Ga/(Ga + In) gradient	η (%)	V_{OC} (mV)	FF (%)	j_{SC} (mA/cm²)	Ga/(Ga + In) integral film composition
A	13.9	636	74.1	29.4	0.303
B	14.9	647	76.9	30.0	0.317
C	13.5	604	72.7	30.7	0.318
D	14.9	640	73.3	31.8	0.289

Measured under AM1.5 illumination at 25°C. The data refer to total area values and are taken from devices without antireflection coating.

in the profiles of Fig. 12.10 with the corresponding optical E_g derived from QE measurements. The values differ by less than 20 meV. Assuming a proper current collection, this result proves that the absorption and the current are determined by the minimum Ga/(Ga + In) in the absorber.

The transition from B to C, i.e., from films with partial Cu-rich growth to pure Cu-poor grown films, is accompanied by a deterioration of V_{OC} and FF. Even the larger gradient with its increased j_{SC} cannot compensate for the growth-related deterioration, resulting in a lower efficiency. We assume

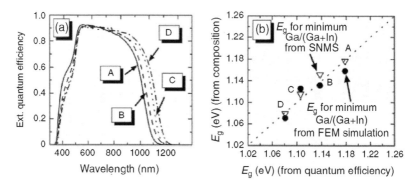

Fig. 12.11. (a) External quantum efficiency QE of devices prepared with processes A–D; (b) Minimum bandgap E_g calculated from minimum Ga/(Ga + In) ratio, obtained from SNMS measurements and FEM simulations, respectively, plotted vs the bandgap energy E_g from QE measurements

that the transition to copper-poor growth is accompanied by an increase in recombination losses, thereby reducing V_{OC} [33].

As already mentioned above, there are different reasons for the beneficial effect of an increasing bandgap towards the back-contact. First, a reduction of back-contact recombination [20] and second, an improved carrier collection by the additional quasielectric field due to the grading of the conduction band [28].

If the diffusion length is larger than the absorber thickness, back-contact recombination becomes important. In this case, the high bandgap at the back-contact may reduce back-contact recombination and therefore improve the cell efficiency. The increased effective diffusion length is less important, as the current collection already is very good in such cells.

If the diffusion length is short compared to the absorber thickness, back-contact recombination is less important; however, the current collection may be improved by the additional quasifield provided by the Ga gradient.

Comparing films with similar growth histories, i.e., A with B (with Cu-rich growth phase) and C with D (without Cu-rich growth phase), the stronger gradient is beneficial for all solar cell parameters. Since the increase in current can be traced back to the decreasing minimum bandgap and since the diffusion length should be quite large in these cells, the improved efficiency might rather be explained by reduced recombination at the back-contact.

Whether or not the current collection can be improved by bandgap grading in the case of an absorber material with short diffusion length, caused, for example, by a high Ga content are questions still to be explored.

12.5 Summary

The development of Cu(In,Ga)Se$_2$ PV modules will reach a point where further improvement is only possible by new innovative concepts. One approach that would increase the efficiency potential of the absorber material due to a better match to the solar spectrum, reduce losses related to the monolithic integration, and improve performance at high temperatures is the increase of the bandgap energy.

We calculated from the diode equation that the conversion efficiency would increase from 16.4% for $E_{\mathrm{g}} = 1.2\,\mathrm{eV}$ to 18% for $E_{\mathrm{g}} = 1.4\,\mathrm{eV}$, under the assumption that the only parameter in Eq. 12.1 and 12.2 to change was E_{g}. Table 12.6 gives an overview of the efficiencies calculated from the diode equation (Sect. 12.1), the losses related to the monolithic integration (Sect. 12.3), and the loss due to a temperature increase from 300 to 325 K (Sect. 12.2) for different bandgap energies. The better match to the solar spectrum and the reduced losses related to the monolithic integration entail a module efficiency of 15.5% at a bandgap energy of $E_{\mathrm{g}} = 1.4\,\mathrm{eV}$ (Fig. 12.8). However, facing the experimentally observed efficiency decline when enhancing the Ga content, such a goal can only be achieved by a concentrated research effort optimizing absorber growth and cell finishing especially for this Ga content in the alloy.

A further approach for improving the module efficiency is realising graded bandgap structures. The "normal grading" is already successfully implemented in an in-line process and leads to efficiency gains of about 7%, whereas the "double grading" approach, for which the highest efficiencies are predicted is difficult to transfer to an in-line process.

Whether or not these approaches are suitable for large area module production will depend on the technical expenses necessary for realising these processes and the yield attainable, as these processes may require a more precise control of process parameters as compared to the relatively forgiving process currently in use.

Table 12.6. Conversion efficiency η calculated from the diode equation (Sect. 12.1), loss due to monolithic integration (Sect. 12.3), and loss due to temperature increase from 300 to 325 K (Sect. 12.2) for different bandgap energies

E_{g} (eV)	η from diode equation (%)	loss due to monolithic integration (%)	loss due to temperature increase from 300 to 325 K (%)
1.2	16.4	15.4	6.7
1.4	18.0	13.4	5.3
1.7	16.6	10.8	4.1

Acknowledgments

The authors acknowledge collaboration with the partners of the Hochspannungsnetz and the financial support by the German Ministry of Research and Education within the Hochspannungsnetz as well as by state of Baden-Württemberg within the PVKonzert project.

References

1. Ramanathan, K., Contreras, M., Perkins, C., Asher, S., Hasoon, F., Keane, J., Young, D., Romero, M., Metzger, W., Noufi, R., Ward, J., Duda, A.: Properties of 19.2% efficiency ZnO/CdS/CuInGaSe$_2$ thin-film solar cells. Prog. Photovolt.: Res. Appl. 11, 225 (2003)
2. Probst, V., Stetter, W., Palm, J., Tölle, R., Visbeck, S., Calwer, H., Niesen, T., Vogt, H., Hernandez, O., Wendel, M., Karg, F.H.: CIGSSE module pilot processing: from fundamental investigations to advanced performance. In: Proceedings of the 3rd World Conference on Photovoltaic Energy Conversion, Osaka, p. 329, 2003
3. Powalla, M., Dimmler, B.: New developments in CIGS thin-film solar cell technology. In: Proceedings of the 3rd World Conference on Photovoltaic Energy Conversion, Osaka, p. 313, 2003
4. Rau U., Electronic properties of wide-gap Cu(In,Ga)(Se,S)$_2$ solar cells. In: Proceedings of the 3rd World Conference on Photovoltaic Energy Conversion, Osaka, vol. 1, pp. 2847–2852 (2003)
5. Gabor, A.M., Tuttle, J.R., Albin, D.S., Contreras, M.A., Noufi, R., Hermann, A.M.: High-efficiency CuIn$_x$Ga$_{1-x}$Se$_2$ solar cells made from (In$_x$,Ga$_{1-x}$)$_2$Se$_3$ precursor films. Appl. Phys. Lett. 65, 198 (1994)
6. Hasoon, F., Yaan, Y., Althani, H., Jones, K.M., Moutinho, H.R., Alleman, J., Al-Jassim, M.M., Noufi, R.: Microstructural properties of Cu(In,Ga)Se$_2$ thin films used in high-efficiency devices. Thin Solid Films 387, 1 (2001)
7. Gabor, A., Tuttle, J.R., Bode, M.H., Franz, A., Tennant, A.L., Contreras, M.A., Noufi, R., Jensen, D.G., Hermannn, A.M.: Band-gap engineering in Cu(In,Ga)Se$_2$ thin films grown from (In,Ga)$_2$Se$_3$ precursors. Sol. Energy Mater. Sol. Cells 4, 247 (1996)
8. Powalla, M., Voorwinden, G., Dimmler, B.: Continuous Cu(In,Ga)Se$_2$ deposition with improved process control. In: Ossenbrink, H.A., Helm, P., Ehmann, H. (eds.) Proceedings of the 14th European Photovoltaic Solar Energy Conference, Barcelona, p. 1270. H. S. Stephens & Associates, Bedford (1997)
9. Rau, U., Jasenek, A., Schock, II.W., Engelhardt, F., Meyer, Th.: Electronic loss mechanisms in chalcopyrite based heterojunction solar cells. Thin Solid Films 361–362, 298 (2000)
10. Malmström, J., Wennerberg, J., Bodegard, M., Stolt, L.: Influence of Ga on the current transport in Cu(In,Ga)Se$_2$ thin film solar cells. In: McNelis, B., Palz, W., Ossenbrink, H.A., Helm, P. (eds.) Proceedings of the 17th European Photovoltaic Solar Energy Conference, p. 1265. WIP-Renewable Energies, München, and ETA, Florence (2002)

11. Kniese, R., Hariskos, D., Voorwinden, G., Rau, U., Powalla, M.: High band gap Cu(In,Ga)S$_2$ solar cells and modules prepared with in-line co-evaporation. Thin Solid Films 431–432, 543 (2003)
12. Turcu, M., Kötschau, I.M., Rau, U.: Band alignments in the Cu(In,Ga)(S,Se)$_2$ alloy system determined from deep level defect energies. Appl. Phys. A 73, 769 (2001)
13. Nishiwaki, S., Dziedzina, M., Schuler, S., Siebentritt, S., Bär, M., Rumberg, A., Rusu, M., Klenk, R., Lux-Steiner, M.C.: Preparation of wide band gap CuGaSe$_2$ solar cells and their optimization. In: Kurokawa, K., Kazmerski, L.L., McNelis, B., Yamaguchi, M., Wronski, C., Sinke, W.C. (eds.) 3rd World Conference on Photovoltaic Solar Energy Conversion, 2003
14. Green, M.A.: Solar Cells—Operating Principles, Technology and System Applications, p. 96. Prentice-Hall, Englewood, NJ (1982)
15. Hanna, G., Jasenek, A., Rau, U., Schock, H.W.: Open circuit voltage limitations in CuIn$_{1-x}$Ga$_x$Se$_2$ solar cells – dependence on alloy composition. Phys. stat. sol. (a) 179, R7 (2000)
16. Shafarman, W.N., Klenk, R., McCandless, B.E.: Device and material characterization of Cu(InGa)Se$_2$ solar cells with increasing band gap. J. Appl. Phys. 79, 7324 (1996)
17. Schmidt, D., Ruckh, M., Grunwald, F., Schock, H.W.: Chalcopyrite/defect chalcopyrite heterojunctions on the basis of CuInSe$_2$. J. Appl. Phys. 73, 2902 (1993)
18. Wada, T., Kohara, N., Nishiwaki, S., Negami, T.: Characterization of the Cu(In,Ga)Se$_2$/Mo interface in CIGS solar cells. Thin Solid Films 387, 118 (2001)
19. Klenk, R.: Characterisation and modelling of chalcopyrite solar cells. Thin Solid Films 387, 135 (2001)
20. Mandelkorn, J., Lamneck, J.: Simplified fabrication of back surface electric field silicon cells and novel characteristics of such cells. In: Proceedings of the 9th IEEE Photovoltaic Specialist Conference, p. 66. IEEE, Silver Springs, NY (1972)
21. Dullweber, T., Lundberg, O., Malmström, J., Bodegard, M., Stolt, L., Rau, U., Schock, H.W., Werner, J.H.: Back surface band gap gradings in Cu(In,Ga)Se$_2$ solar cells. Thin Solid Films 387, 11 (2001)
22. Kötschau, I., Kerber, H., Wiesner, H., Hanna, G., Schock, H.W.: Band gap grading in Cu(In,Ga)(S,Se)$_2$-based solar cells. In: Scheer, H., McNelis, B., Palz, W., Ossenbrink, H.A., Helm, P. (eds.) Proceedings of the 16th European Photovoltaic Solar Energy Conference, Glasgow, p. 724. James & James Ltd, London (2000)
23. Turcu, M., Rau, U.: Fermi level pinning at CdS/Cu(In,Ga)(Se,S)$_2$ interfaces: effect of chalcopyrite alloy composition. J. Phys. Chem. Solids 64, 1591 (2003)
24. Turcu, M., Pakma, O., Rau, U.: Interdependence of absorber composition and recombination mechanism in Cu(In,Ga)(Se,S)$_2$ heterojunction solar cells. Appl. Phys. Lett. 80, 2598 (2002)
25. Huang, C.H., Sheng, S., Li, S., Anderson, T.J.: Device modeling and simulation of CIS-based solar cells. In: Proceedings of the 29th IEEE Photovoltaic Specialist Conference, New Orleans, p. 748. IEEE, New York (2002)
26. Topic, M., Smole, F., Furlan, J.: Band-gap engineering in CdS/Cu(In,Ga)Se$_2$ solar cells. J. Appl. Phys. 79, 8537 (1996)
27. Dullweber, T., Rau, U., Contreras, M.A., Noufi, R., Schock, H.W.: Photogeneration and carrier recombination in graded gap Cu(In,Ga)Se$_2$ solar cells. IEEE Trans. Electron. Devices ED47, 2249 (2000)

28. Urli, N.B., Desnica, U.V., Coffou, E.: Collection efficiency of drift-field photode-tectors made from low-lifetime materials. IEEE Trans. Electron. Devices ED32, 1077 (1975)
29. Sinkkonen, J.: A simple model of a graded band-gap pn-heterojunction cell. In: Proceedings of the 1st World Conference on Photovoltaic Energy Conversion, Hawaii, p. 1826, 1994
30. Voorwinden, G., Powalla, M.: In-line Cu(In,Ga)Se$_2$ co-evaporation on 30 × 30 cm^2 substrates. Thin Solid Films 387, 37 (2001)
31. Voorwinden, G., Powalla, M.: Modeling Cu(In,Ga)Se$_2$ growth conditions for In-line co-evaporation processes. In: McNelis, B., Palz, W., Ossenbrink, H.A., Helm, P. (eds.) Proceedings of the 17th European Photovoltaic Solar Energy Conference, Munich, p. 1203. WIP-Renewable Energies, München, and ETA, Florence (2002)
32. Dimmler, D., Dittrich, H., Menner, R., Schock, H.W.: Performance and opti-mization of heterojunctions based on Cu(Ga,In)Se$_2$. In: Proceedings of 19th IEEE Photovoltaic Specialist Conference, New Orleans, p. 1454. IEEE, New York (1987)
33. Rau, U., Schock, H.W.: Electronic properties of Cu(In,Ga)Se$_2$ solar cells – recent achievements, current understanding, and future challenges. Appl. Phys. A 69, 131 (1999)

Index

Printed by Publishers' Graphics LLC
MLSI130417.10.24.80